Dismantling the Bomb and Managing the Nuclear Materials

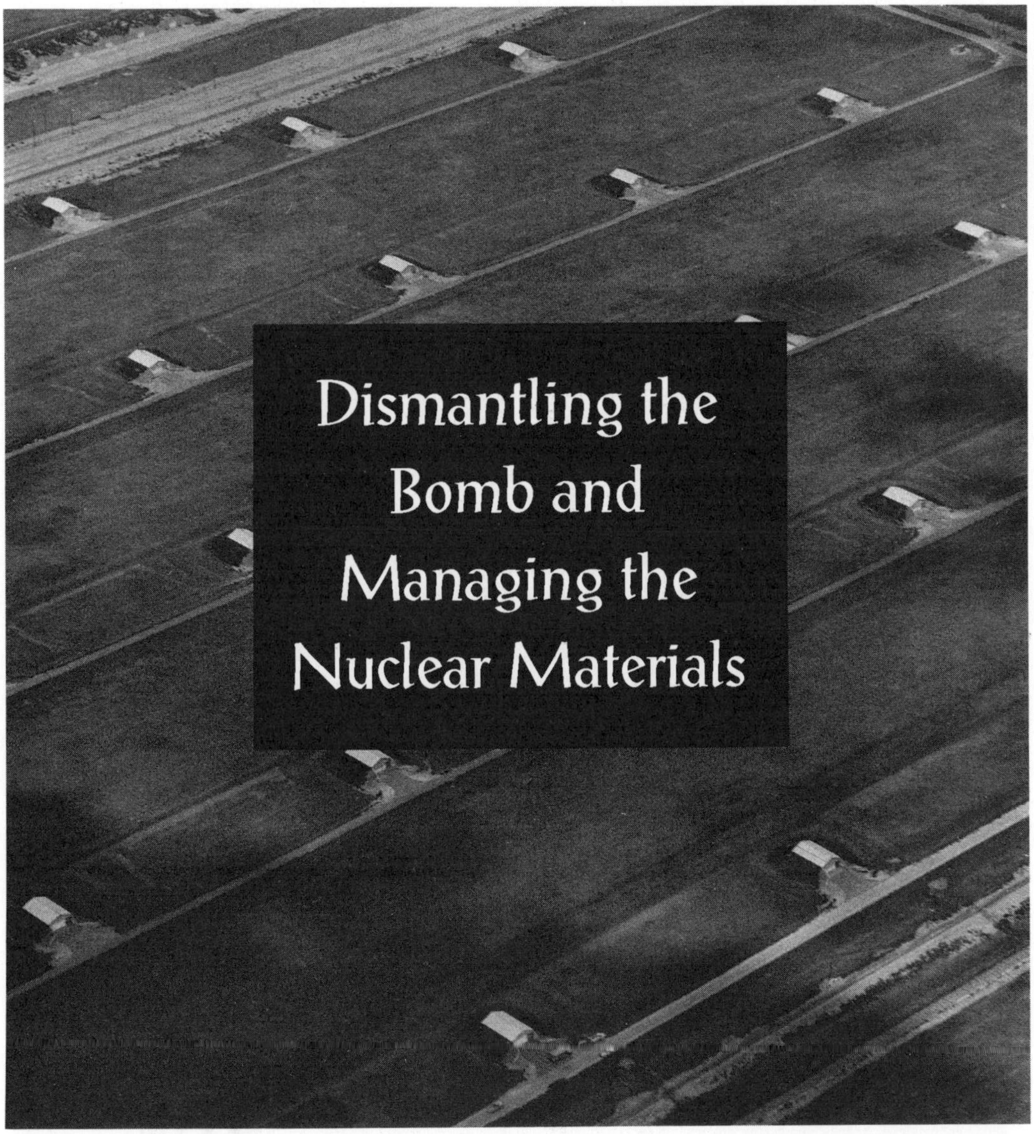

PHOTO BY ROBERT DEL TREDICI

Nuclear warhead storage bunkers

U.S. Congress

Office of Technology Assessment

Recommended Citation:
U.S. Congress, Office of Technology Assessment, *Dismantling the Bomb and Managing the Nuclear Materials,* OTA-O-572 (Washington, DC: U.S. Government Printing Office, September 1993).

For sale by the U.S. Government Printing Office
Superintendent of Documents, Mail Stop: SSOP, Washington, DC 20402-9328
ISBN 0-16-041968-9

Foreword

The Nation has embarked on a bold new mission to enhance world peace through deep and lasting cuts in nuclear arsenals. It has removed thousands of nuclear weapons from active, deployed status and has begun eliminating delivery systems and dismantling the warheads themselves. Our old Cold War adversary, the former Soviet Union, has taken similar steps. The United States and Russia have pledged to continue their programs of weapons retirement and warhead dismantlement, and are discussing methods of defining and achieving long-term goals.

Such efforts are unprecedented and challenging; they require resources and talent as well as enduring dedication within Government institutions. People are concerned that the work be conducted so as to avoid the types of adverse environmental and health impacts that resulted from nuclear weapons production in the past. Experts have been investigating how to use, control, or dispose of the nuclear materials—plutonium and highly enriched uranium—recovered from dismantled warheads. For these reasons, the Senate Committee on Governmental Affairs requested that the Office of Technology Assessment conduct a study of the key technical, policy, and institutional options to be considered in the Federal Government's plans. The Senate Committee on Foreign Relations endorsed this request. This report presents the results of OTA's investigations and analyses.

Although current Federal efforts are adequate for the present, they are insufficient to meet the long-term challenge ahead. OTA concludes that the success of future warhead dismantlement and materials management requires a focused, high-level governmental effort to develop a comprehensive national policy. It also requires an open decisionmaking process and capable institutions to set and implement long-range goals and plans. In this report, OTA suggests various initiatives that Congress could consider to establish a national policy, determine the next steps in warhead dismantlement and nuclear materials management, approach decisions on the ultimate disposition of nuclear materials, enhance the institutional capabilities necessary to ensure success, and encourage sound dismantlement and materials management in Russia.

Substantial assistance was received from many organizations and individuals for this study. OTA sincerely appreciates the guidance received from its advisory panel, workshop participants, numerous reviewers, contributors, consultants, and contractors. We also received help from several Federal agencies, including the Departments of Energy, Defense, and State, and other Federal, State, and local agencies. Without this cooperation and expert advice, OTA would not have been able to accomplish this study.

Roger C. Herdman, Director

Advisory Panel

John E. Till
Chairman
Radiological Assessments Corp.

John F. Ahearne
Sigma Xi, The Scientific Research Society

Charles Barrett
Oil, Chemical, and Atomic Worker's International Union

Thomas A. Burke
School of Hygiene and Public Health
The Johns Hopkins University

Beverly E.C. Gattis
STAND of Amarillo, Inc.

Sherri Wasserman Goodman, Esq.[1]
Goodwin, Procter and Hoar

Kenneth Lichtenstein
Division of Infectious Disease
Rose Medical Center

Robert R. Loux
Nuclear Waste Project Office
State of Nevada

Daniel S. Miller, Esq.
Colorado Office of the Attorney General

John M. Napier
Techniques, Inc.

Robert H. Neill
Environmental Evaluation Group
State of New Mexico

Louis Peoples
Madison Gas & Electric Co.

Ambassador Rozanne L. Ridgway
The Atlantic Council of the United States

Frank N. von Hippel
Woodrow Wilson School of Public and International Affairs
Princeton University

James D. Werner[1]
Natural Resources Defense Council

John M. Whiteley
Professor, Social Ecology and Social Sciences
University of California at Irvine

NOTE: OTA appreciates and is grateful for the valuable assistance and thoughtful critiques provided by the advisory panel members. The panel does not, however, necessarily approve, disapprove, or endorse this report. OTA assumes full responsibility for the report and the accuracy of its contents.

[1]James D. Werner and Sherri Wasserman Goodman served on the Panel before joining DOE and DOD, respectively.

Project Staff

Clyde Behney
Assistant Director
Health, Life Sciences and the Environment

John Andelin
Assistant Director, to Aug. 31, 1993

Robert Niblock
Program Manager
Oceans and Environment

PETER A. JOHNSON
Project Director

Emilia L. Govan
Project Director

Mark Brown
Senior Analyst

Joan Harn
Senior Analyst

German Reyes
Analyst

Kathy Cox
Analyst

Paul Carroll
Research Analyst

ADMINISTRATIVE STAFF

Kathleen Beil
Office Administrator

Kimberly Holmlund
Administrative Secretary

Sharon Knarvik
Secretary

CONTRIBUTORS

Isaac Amuah
Jan Linsenmeyer
Allison Lowery
Tara O'Toole

CONTRACTORS

Jackie Braitman
Kahtilya Consulting, Inc.

Robert Budnitz
Future Resources Associates

Stan Kolar
Kolar Associates

Stephen Lipmann
Environmental Research and Analysis

Arjun Makhijani
Institute for Energy and Environmental Research

Robert Morgan
Washington University

Florence Poillon
Editor

Monica Schoch-Spana
Consultant

Workshop Participants

PLUTONIUM DISPOSITION WORKSHOP

John Ahearne
Sigma Xi

Carl Anderson
National Research Council

Robert Bernero
Nuclear Regulatory Commission

Carl Bloomster
Pacific Northwest Laboratory

Robert Civiak
Office of Management and Budget

Thomas Cochran
Natural Resources Defense Council

Brian Costner
Energy Research Foundation

Alan Croff
Oak Ridge National Laboratory

Harold Feiveson
Princeton University

Thomas Fox
Pacific Northwest Laboratory

Arjun Makhijani
Institute for Energy and Environmental Research

Marvin Miller
Massachusetts Institute of Technology

Loring Mills
Edison Electric Institute

William Sutcliffe
Lawrence Livermore National Laboratory

Ted Taylor
Consultant

Louis Willett
U.S. Department of Energy

PLUTONIUM STORAGE ROUNDTABLE

Howard Canter
U.S. Department of Energy

Jesse Cleveland
U.S. Geological Survey

Paul Cunningham
Los Alamos National Laboratory

Ted Dobry
U.S. Department of Energy

Joel Gingold
S.M. Stoller Corp.

Charles Haughney
Nuclear Regulatory Commission

Dewey Large
Scientific Ecology Inc.

Paul Maddox
Westinghouse Savannah River Corp.

Ed Moore
Westinghouse Savannah River Corp.

Duane Schmoker
Pacific Nuclear Corp.

William Shuler
U.S. Department of Defense

Victor Trelis
U.S. Department of Energy

John Trout
U.S. Corps of Engineers

Louis Willett
U.S. Department of Energy

Contents

APPENDIXES

Summary 1

During the Cold War, both the United States and the Soviet Union built and maintained large stockpiles of nuclear weapons. Over the past 2 years, the leaders of these nations have pledged to withdraw tactical weapons and sharply reduce the size of the strategic weapons arsenal. Both nations have begun to retire thousands of weapons and to dismantle the nuclear warheads—the part of the weapon that contains its massive destructive power. Reducing the nuclear arsenals of both nations presents a unique opportunity and a challenge. The opportunity is to eliminate large numbers of warheads and reduce the threat of nuclear war. The challenge is to devise feasible and practical means of dismantling them and managing the constituent nuclear materials without causing new environmental, safety, or security problems. Still needed are decisions, policies, and plans to guide both the short- and the long-term goals of this effort.

Treaty agreements, such as the Intermediate Range Nuclear Forces (INF) and the Strategic Arms Reduction (START) treaties, negotiated to date require that weapons be retired from deployed status and that the means of delivering them be removed or destroyed. They do not require that warheads be dismantled or that warhead parts and materials be destroyed. However, the United States has undertaken to remove certain weapons from the stockpile, return warheads to the facilities that assembled them, dismantle the warheads, and store or dispose of their components, parts, and key nuclear materials. Substantial disassembly work is ongoing. The specific plans and schedules, however, are not available to the general public. Nor is the ultimate scope of this effort.

"Successful dismantlement and disposition of the weapons materials may be the single most important public health, environmental, and social challenge we face."

Public health expert participating at OTA panel meeting

"Current dismantlement can either be done well and set a foundation for future progress, or it can be done badly, leaving so much unaccounted for, so much room for uncertainty, so much inequity that it will set back, if not destroy, future possibilities."

Local citizen's group reviewer of OTA report

The Office of Technology Assessment (OTA) has analyzed the present U.S. approach to this undertaking and concludes that current Federal efforts are insufficient to the challenge because they are scattered and lack uniform objectives; they are not based on a clear mission; the public distrusts the responsible Federal agencies, and fears that the environmental and health impacts may be no better than past performance; and there has been little informed public debate to establish national goals. In essence, the Nation has no coordinated, comprehensive national policy on nuclear warhead dismantlement, and current overall management of the task is weak.

Neither the United States nor Russia has developed a technically and politically feasible plan to dismantle warheads and dispose of the nuclear materials from them. Policies for nuclear warhead dismantlement and materials control are important to both U.S. and international security. While recent pronouncements and agreements by national leaders may set goals for reduction of the weapons stockpile, they do not, by themselves, eliminate nuclear warheads. Although nuclear weapons can be rendered less threatening by destroying the means of delivering them (as recently negotiated treaties require), destroying warheads and their constituent nuclear materials safely and effectively is a very difficult task. Many of the most dangerous materials will need careful management for generations.

OTA's analysis of the dismantlement program makes clear that eliminating these warheads—or even destroying a portion of the stockpile of nuclear weapons that have been amassed—will be neither simple nor painless. The difficulties of weapons retirement and warhead dismantlement should not be underestimated. Plans for long-term storage or disposition of nuclear materials must be resolved, and difficult decisions regarding these matters must be made at the highest levels of government.

THE CHALLENGE

Tens of thousands of nuclear weapons are still deployed in the United States, Russia, and other nations (i.e., ready for use or deliverable). Others, although not deployed, are part of what is called —in the United States—the "reserve" stockpile, meaning they are maintained as "backups" for deployed weapons. Still other weapons are removed from both the active stockpile and the inactive reserve, and "retired." The warhead portions of the retired weapons are eventually returned to a Weapons Complex plant for dismantlement.[1]

The Strategic Arms Reduction Treaty, START II, which awaits ratification, provides for some warheads that are presently deployed to be separated from delivery vehicles or otherwise placed in a status in which they are not deliverable or ready for use. START II does not impose any requirements to actually dismantle the warheads that are removed from deployed status. Neither START agreement calls for dismantling any warheads that are now in the reserve stockpile or that may be added to it in the future.

Potential political instability in the former Soviet Union raises concern that control over some weapons will diminish and they will fall into the hands of revolutionary regimes or terrorist groups. The potential proliferation of nuclear weapons poses a serious threat to international security. There is also the possibility that a weapon may detonate accidentally or pose other types of safety problems. Accidental explosions are a concern if groups with limited technical capability and resources have control of these weapons.

[1] Dismantlement means the removal of all nonnuclear components, including the chemical high explosive that surrounds the nuclear materials. Dismantlement also includes waste management and disposal of other parts and materials. It does not, however, include destruction of the key nuclear materials or even of the major nuclear subassemblies.

For these and other reasons, the criteria against which options for dismantlement, storage, and disposition of components from nuclear warheads must be assessed differ significantly from those that applied to warhead production. In the past, national security was accorded much more prominent attention than environmental risks. Today, however, there is a need for responsible stewardship of the long-lived nuclear materials that are bequeathed to future generations, and safeguards will be required to protect the safety and health of the public and of the workers who carry out dismantlement.

New technologies may offer solutions or partial solutions to some of the problems associated with either safe disposal or utilization of materials whose radioactive half-lives are measured in many thousands of years. Yet few proven technologies are readily available. Nonproliferation concerns will affect decisions about technologies because of the pressure to come up with options that reduce the risks of nuclear materials being easily diverted into new warheads.

Dismantlement of nuclear warheads is proceeding at a time when trust of government in general, and DOE in particular, is--at best--fragile. The culture of secrecy and insularity embraced by the Department of Energy (DOE) and its predecessor agencies has had a corrosive effect on relations between the Department and the communities neighboring nuclear weapons facilities. The United States begins with the handicap of widespread public mistrust of its own institutions charged with these responsibilities because of their previous failures to safeguard the environment and health. Thus, one of the first tasks is to rebuild institutional credibility.

To do this, the priorities and characteristics of the institutions that supported warhead production will have to be carefully rethought. Greater attention to environmental, safety, and health impacts is essential. If the United States is to successfully carry out nuclear warhead dismantlement and materials management and disposition, and to engage in cooperative efforts with Russia, new institutional capabilities and management approaches are essential. These institutions will be expected to devote much more attention to the environmental impacts of proposed ways of handling nuclear materials than was given when warhead production was the primary concern.

THE OFFICE OF TECHNOLOGY ASSESSMENT STUDY

This study addresses the challenge of eliminating thousands of nuclear warheads. It traces the U.S. process within the responsible Federal agencies, with particular attention to factors that may affect realization of the national goal of safe and secure stockpile reduction in a manner that protects human health and the environment. The report also reviews related work in Russia, focusing on the ability of the United States to influence a safe, secure, and environmentally sound process there.

If the United States wishes to develop and implement policies leading to substantial nuclear arms reduction worldwide, as well as to substantial reduction of the nuclear materials with which to make new warheads, certain actions are important and probably more urgent than generally realized. This report discusses the following major activities involved in the unprecedented enterprise to achieve nuclear stockpile reduction:

- the process for retiring weapons from active deployment in the military and returning their nuclear warheads to the facilities that manufactured them; and dismantlement of the warheads, and subsequent handling of the parts and materials from them; and
- the storage, control, and ultimate disposition of key nuclear materials (plutonium and highly enriched uranium) from the warheads.

Box 1-A—Key Findings

- Ongoing Federal programs and plans within the Departments of Defense and Energy for retirement and dismantlement of nuclear weapons are currently treated as a short-term modification of existing practice rather than a change in focus from past missions of production and stockpile maintenance.
- Numbers of weapons in the active stockpile, and numbers to be retired and dismantled, are contained in classified documents not available to the general public. Existing and pending international agreements do not require that any warheads be dismantled, only that they be removed from delivery systems. The nation's massive nuclear stockpile is now partly dismantled, partly in temporary storage, partly in transition, and partly deployed.
- Environmental, safety, and health problems continue in the operation of the DOE Weapons Complex, and certain aspects of current dismantlement activities—the use of old facilities, additional sources and generation of waste, and slow adoption of modern health and safety practices—may affect the success of dismantlement programs.
- A continuing lack of public credibility may have a major impact on progress in dismantlement and on implementing key operational decisions. Public interest groups have obtained a legitimate voice in influencing DOE operations through environmental legislation and their political power. Despite new public participation initiatives, the major DOE sites have yet to ensure adequate communication with the public, to understand public concerns, or to involve the public in critical decisions.
- It is likely that significant portions of the highly enriched uranium and plutonium recovered from dismantled warheads will need to be stored for decades regardless of the ultimate disposition option chosen for them. Significant time will be required for making disposition decisions and formulating policies; for planning, designing, funding, building, and testing even the most available technology; for gaining regulatory and public acceptance; and for actually processing quantities of materials.
- The use of surplus plutonium from weapons as fuel for U.S. commercial reactors is unlikely because of economic factors, the concerns of U.S. utilities about regulatory constraints and public acceptance, and the need to evaluate U.S. policies that discourage commercial plutonium use.
- If the policies articulated urge expeditious processing or conversion of plutonium to less weapons-usable forms, it may be best to pursue the most available near-term technologies. OTA finds that a process to immobilize it directly in some form such as vitrified glass or, with appropriate poisons, to decrease its proliferation risk and a Government-built and operated dedicated light-water reactor that uses mixed plutonium and uranium fuels are two such near-term technologies.
- It is impossible to fission plutonium completely (and thus "destroy" all of it), but certain new developments may be able to convert it to different radionuclides at a much more efficient rate than existing technologies. However, the research required to develop such advanced reactors and converters would be costly, and would require times on the order of decades.
- The U.S. program to assist Russia with nuclear warhead dismantlement has initiated important cooperative work but has not addressed the broader issues of mutual goals and interests in stockpile and materials reduction or control, nor has it had a significant effect on Russian dismantlement.
- The United States has not verified specific warhead dismantlement activities and accomplishments in Russia, and has no direct cooperative process for developing accurate information about Russian dismantlement status and capabilities.
- Efforts to integrate U.S. warhead dismantlement plans with programs to assist Russia have not received substantive attention. There is little linkage between Russian economic, environmental, or social needs and U.S. programs to assist and encourage Russian dismantlement or related activities.
- While the United States views expeditious Russian warhead dismantlement and materials disposition as vital to its national security, Russia's agenda is dominated by economic and political issues that could relegate dismantlement to a low priority.

The following sections summarize the status of ongoing dismantlement activities, OTA's findings, and an analysis of the policy issues involved and the initiatives proposed.[2] The key OTA findings listed in box 1-A summarize the major points discussed in this report. The findings address U.S. warhead dismantlement and materials management, and U.S. cooperation with Russia regarding the disposition of weapons in the former Soviet Union.

Box 1-B lists issues related to the process of nuclear warhead dismantlement and materials management. These issues are presented in the form of questions and relate to the major decisions that the United States will have to make to facilitate dismantlement both here and in the former Soviet Union.

Finally, box 1-C presents the key policy initiatives developed by OTA in this report. These initiatives are intended to offer possible approaches to improve Government programs and enhance their chances of success. They could be adopted either through legislative initiatives or by the Administration with congressional encouragement. The options can be pursued either individually or as a group. They are presented in the order in which they are discussed in chapter 7.

DISMANTLEMENT OF NUCLEAR WARHEADS

Status

According to a long-standing administrative procedure for management and control of nuclear weapons within Federal agencies and the military services, dismantlement begins with a presidential decision approving the annual Nuclear Weapons Stockpile Plan. Retired weapons are then transferred to a military base within the continental United States, where the warhead is usually separated from the delivery system and returned to the DOE facility that assembled it. DOE retains custody until it is dismantled and its components have been disposed of. In recent years, thousands of U.S. nuclear weapons have been put on retirement status: many of these have been returned to DOE for dismantlement; others are in storage at military bases, waiting their turn in the dismantlement process. In FY 1993, the United States expects to dismantle about 1,400 warheads, but plans for the total number of weapons to be retired and disassembled, as well as the future size of a reduced warhead stockpile, are not available for public release.

Warheads returned from the Department of Defense (DOD) to the Department of Energy for dismantlement are transported to the DOE Pantex Plant near Amarillo, Texas, where they were built. Several Department facilities are currently engaged in warhead dismantlement and related work, with major activities centered at Pantex, Y-12 (in Oak Ridge, Tennessee), and Savannah River (in Aiken, South Carolina). At Pantex, plutonium pits (the primary explosive parts) are removed from warheads, placed in containers, and stored in bunkers. Other parts and wastes are characterized, stored, and disposed of in a variety of ways. Nuclear warhead "secondaries" and highly enriched uranium (HEU) are shipped to the Y-12 Plant at Oak Ridge for further storage or disassembly. Tritium gas canisters are shipped to the Savannah River Plant for storage or processing.

The United States has recently announced that it will no longer produce weapons-grade plutonium or highly enriched uranium for warheads. In practice, these activities ceased some years ago, and production facilities have not been operating. Thus, the United States plans to store some of the materials extracted from disassembled warheads

[2] The analyses in this report are based on unclassified information. Thus, certain data such as weapons types, numbers of weapons, retirement schedules, warhead designs, materials shapes, and some processes are discussed only in general terms. OTA did have access to classified information in the course of the study and has prepared a classified annex to this report, which contains more detailed information regarding the nuclear weapons stockpile, future plans with respect to nuclear weapons, and related data.

Box 1-B—Issues Related to Weapons Dismantlement and Materials Management

U.S. WEAPONS DISMANTLEMENT AND MATERIALS MANAGEMENT

Policy and Strategy

How many U.S. warheads are to be retired and dismantled?

How much weapons-grade plutonium and highly enriched uranium (HEU) from already dismantled weapons and from weapons planned for dismantlement will not be required for stockpile purposes, and can thus be declared surplus?

Should information about numbers of weapons to be dismantled and amounts of surplus materials from dismantled weapons be made public?

Should surplus plutonium pits from U.S. warheads be stored indefinitely or disposed of as waste?

Should surplus HEU from U.S. warheads be stored indefinitely or converted for use in commercial power reactors?

Should U.S. surplus materials be made amenable to monitoring or inspection under a bilateral arrangement with Russia?

Operations and Management

When will dismantlement of retired weapons or weapons planned to be retired be completed?

What additional measures should be taken to manage the dismantlement mission so as to protect the environment, as well as public and worker health and safety?

How long should plutonium pits from dismantled warheads be retained in temporary storage at Pantex—and the HEU from dismantled warheads in storage at Y-12?

What type of processing facility is needed to maintain the plutonium pits?

What type of facilities are needed for long-term storage of plutonium and HEU (pending some future use or disposal), and where could such facilities be located?

What type of technologies should be used if plutonium is deemed to be a waste, and what facilities are needed to implement disposal plans?

Should the surplus materials from dismantled warheads be stored separately from materials needed for weapons stockpile requirements?

To what extent can and should operational information be made available to the public, and how can public participation best be ensured?

Through what process will a site or sites be chosen for facilities required to carry out ultimate disposition options including long-term storage, conversion to fuel, or disposal as waste?

Organizational Structure

Should responsibility for management and disposition of surplus materials from warheads be retained in the Department of Energy's Defense Programs, or given to a new organization within DOE or another existing agency, or should a new organization be created for this purpose?

How should the transition be made between the present organizational structure and a potential future one?

How can external oversight and enforcement be strengthened—what agencies should be engaged, and what mechanisms should be developed?

> **RUSSIA'S NUCLEAR WEAPONS DISMANTLEMENT AND MATERIALS MANAGEMENT**
>
> How can the United States best encourage and aid Russia in dismantling warheads, and in the management and disposition of materials from them, and how should those efforts be structured?
>
> Should the United States propose or enter into reciprocal arrangements with Russia involving information exchange, transparency, and inspections?
>
> Should the United States encourage or promote any role by an international organization with respect to Russian weapons and nuclear materials?
>
> Should the United States enter into joint study projects or provide technical assistance to Russia for processes leading to ultimate disposition of plutonium?

for possible future military use. The facilities that were used to recycle old warhead parts such as plutonium pits have been shut down, largely for environmental and safety reasons.

Plutonium pits from recently dismantled warheads are being stored at the Pantex Plant, where warhead disassembly takes place. DOE is running out of storage space for plutonium pits at Pantex and wants to change the storage configuration in existing bunkers to accommodate more pits, but the specific plan has not been approved yet. HEU from disassembled warheads is now being stored at the Y-12 facility, and there are no current plans to store it elsewhere.

Findings

The Nation's massive nuclear stockpile is now partly dismantled, partly in temporary storage, partly in transition, and partly deployed. Whereas past dismantlement activities were geared to maintaining the weapons stockpile, present and future activities are intended to permanently reduce it. Since fewer new weapons will be made, most of the materials recovered from dismantled warheads will no longer be recycled for use in other weapons. More plutonium and HEU will have to be stored and managed for long periods of time, and international factors may have significant impacts on materials management decisions. Yet, Federal programs and plans within DOD and DOE for retirement and dismantlement of nuclear weapons are currently treated as a short-term modification of existing practice, rather than a change in focus from the past missions of production and stockpile maintenance.

Existing and pending international agreements require only the removal of warheads from delivery systems. Preparation for long-term institutional custody of warheads and their nuclear materials lacks direction. DOE does not have comprehensive and accurate estimates of the total current or future annual costs of this enterprise, but available information indicates that DOE expenditures for dismantlement activities at all sites could be approaching $1 billion annually.

Thus far, there have been few if any serious problems with respect to dismantlement, but some process difficulties and logistical problems have caused schedule changes. One potential stumbling block is the storage of plutonium pits from warheads. Although DOE has stated that it needs to change the storage configuration in its World War II-vintage bunkers at Pantex to accommodate the anticipated number of pits coming from warheads, it has not yet produced the documentation required for approval. The State of Texas, community groups, and other experts have found DOE's environmental analysis to be deficient and have objected to the fact that DOE originally restricted access to the associated safety review. In addition, some citizen groups in Texas are concerned that although DOE says the pits will remain in "temporary storage" for 6 to 10 years, Pantex could turn into a de facto long-term storage site, and the pits may

Box 1-C—Summary of Policy Initiatives

Congress could implement—or the Administration could undertake to implement—the following policy initiatives:

Initiative 1—A National Dismantlement Policy

Develop and announce a national policy that sets goals for warhead dismantment and materials management, and specifies the amount of plutonium and highly enriched uranium from dismantd warheads that will not be needed to support future stockpile requirements.

Initiative 2—Strengthening DOE Management

Implement a DOE management system that gives priority to protecting the environment, health, and safety; expand and strengthen external oversight of DOE dismantment and materials management activities by independent outside entities.

Initiative 3—Nuclear Materials Storage

Establish an interagency task force that includes Federal agencies with expertise in regulatory, international, and public involvement matters to recommend a plan for safe, secure storage of nuclear materials, and to develop a process acceptable to the interested public for siting new or modified storage facilities.

Initiative 4—Nuclear Materials Disposition

Create a national commission to recommend goals, policies, and programs for ultimate disposition of surplus plutonium and HEU from warheads, and to provide a basis for developing an ultimate disposition policy for these materials.

Initiative 5—A New Materials Management Organization

Create a new organization outside DOE to manage surplus materials from warheads, or establish a new organization for this purpose within DOE or some other existing agency.

Initiative 6—Information Access

Review and possibly revise the existing legal basis for restricting access to information in light of today's post-Cold War national security objectives, and accelerate efforts to increase access to information relevant to warhead dismantment and materials disposition.

Initiative 7—Cooperation with Russia

Strengthen the relationship between U.S. assistance to Russia for materials disposition and other programs in which assistance is desired by Russia; develop a means for joint assessment of plutonium disposition technologies; and negotiate mutual disclosure of information and reciprocal materials monitoring arrangements.

deteriorate before alternative storage arrangements are available.

Another stumbling block is DOE's poor record with respect to environmental and safety matters at its Nuclear Weapons Complex in the past, which has led to concerns among the interested public and affected communities about future DOE activities at those sites. Recent process difficulties during dismantlement at Pantex have caused the public to continue to question health and safety practices. Lack of public trust and credibility could adversely affect prospects for successful conduct of dismantlement and materials management activities.

While DOE is working on improvements to its environmental, health, and safety programs at the

Dismantlement means more than putting weapons under wraps like these "extinct" bombs at the National Atomic Museum in Albuquerque.

Nuclear Weapons Complex, current dismantlement activities still face problems, such as the use of old facilities, waste generation, and the slow adoption of modern worker health and safety practices. DOE's lack of public credibility could also have a negative impact on prospects for making key operational decisions regarding dismantlement and management of materials from warheads. Despite DOE's efforts to develop better public participation initiatives, the major dismantlement sites have yet to ensure adequate communications with the public, address public concerns, or involve the public in making decisions about dismantlement and materials management that could affect surrounding communities. In addition, considerable work remains to develop a national consensus around dismantlement goals and to ensure the protection of human safety, environmental integrity, and international security.

Policy Issues and Initiatives

Although present efforts to dismantle warheads and manage warhead materials are being treated by DOD and DOE as business as usual, these activities should be viewed as constituting a new mission with different challenges than in the past. Failure to effectively carry out the new mission here could adversely affect similar efforts abroad, with harmful consequences for international security and the global environment.

A NATIONAL DISMANTLEMENT POLICY

To define the new mission, and guide the agencies in implementing it, the Nation could establish a policy that sets forth the long-term goals and rationale for dismantlement. As part of that policy, decisions about the number of weapons to be retired and dismantled, as well as the time frame for dismantlement, would be made public. The Administration will also have to decide on the amount of plutonium and HEU currently available from dismantled warheads that is not needed to support nuclear weapons stockpile requirements and could be declared surplus to military needs. To aid this process, Congress could direct that an unclassified report containing such information be prepared and updated annually. This initiative would facilitate understanding of the rationale and goals of dismantlement; help ensure the public that future actions are consistent both with safety and protection of human health and the environment, and with U.S. strategic needs; and signal the international community that the United States is serious in its intent to dismantle warheads.

STRENGTHENING DOE MANAGEMENT

Although DOE is attempting to establish new guidelines for protecting the environment, health, and safety in its dismantlement and nuclear materials management activities, these matters require continuing attention. It is critical for DOE to develop a management system at all levels of its organization that is strongly committed to environmental, safety, and health improvements, and that effectively integrates this commitment into its operations. To help ensure that this occurs, external oversight of DOE's dismantlement and materials management program and plans should be strengthened. One way to accomplish this is for Congress to provide the Defense Nuclear Facilities Safety Board with the necessary resources and personnel (and any additional authority required) for this purpose. To assure communities around the sites that activities are being conducted properly, the Board—as well as DOE—could provide greater opportunity for public involvement than in the past. In addition, the Occupational Safety and Health Administration (OSHA) could be given jurisdiction over DOE worker health and safety.

In general, Congress could insist that DOE upgrade and strengthen its management systems to adopt and maintain high standards of worker health and safety, public health, and environmental protection.

ROBERT DEL TREDICI

Models of World War II nuclear weapons Fat Man and Little Boy at the Bradbury Science Museum in Los Alamos, New Mexico. Conventional explosive hemispheres that surround the plutonium pit in a nuclear warhead are shown in the foreground.

MANAGEMENT OF NUCLEAR MATERIALS

Status

The two principal nuclear materials in warheads are plutonium and highly enriched uranium. Together or separately, they can be made into new warheads; thus there is a need to keep these materials safe and secured. Because of their radioactive half-lives, these materials will continue to pose some level of risk to human health and the environment for many thousands of years. OTA has thus focused on plutonium and HEU, although the disposition and disposal of many other materials from dismantled warheads are of concern.

A few hundred tons of plutonium and more than a thousand tons of HEU (exact numbers are classified) were produced worldwide for warheads. Today, this stockpile exists either in intact warheads or weapons, in forms ready to be made into warheads, or as pits and other forms removed from retired weapons. The United States and Russia have by far the largest portion of these materials. Both plutonium and uranium are also found in various forms and quantities in the nuclear industry worldwide, along with other industries that use nuclear materials. Some weapons-grade HEU is used in naval and research reactor fuel. Some plutonium that has been separated from commercial spent fuel could also be used in warheads even though it was not made for such use.

Nuclear materials taken from dismantled U.S. warheads, including plutonium pits placed in the bunkers at the Pantex Plant and HEU housed at the Oak Ridge Y-12 Plant, are considered to be in temporary or interim storage. Although DOE has stated its intention to store plutonium pits in temporary bunkers at Pantex for the next 6 to 10 years, it has not announced any plans to provide safe storage for the pits beyond that time. DOE also has not indicated its long-term storage plans for HEU.

Long-term or permanent solutions to the disposition[3] of these materials await policy decisions by the President and Congress. DOE has not declared any of this material to be surplus. However, recent DOE-sponsored studies have

[3] In this report, the term "disposition" means the spectrum of possibilities about what to do with these materials beyond weaponry—first to store them in a safe, secure facility; perhaps to destroy some portions if technically feasible and practical; perhaps to utilize them to produce civilian energy, if security is adequate and if the technology and economics prove sound; and finally to dispose of them as waste if technology and national policies permit.

focused on options for plutonium disposition through the use of various fuels containing plutonium in existing or advanced nuclear reactors. Within these reactors the plutonium would be irradiated and some of it converted to other radionuclides and fission products, possibly with the generation of electricity. Other studies have addressed plutonium storage for moderate to long-range time frames and techniques for turning plutonium into a form suitable for disposal as waste. Debate over these options is based largely on whether plutonium is viewed as a valuable asset whose beneficial uses are to be explored or a major liability to be disposed of in the safest and most secure way.

It is extremely difficult to convert significant amounts of plutonium into a substance that would be nonradioactive or harmless to health and the environment. Existing reactor technologies can be used to consume it as fuel, that is, to irradiate it and transform portions of it over time. Mixed-oxide (plutonium and uranium) fueled reactors are an example of existing technology that may be modified or adapted for plutonium disposition. Advanced reactor or converter technologies could be developed to achieve a large degree of plutonium transformation (and perhaps also to produce energy). However, available information indicates that their development would require significant time and resources, and it is uncertain how effective they would be. Alternatively plutonium could be disposed of more directly by using available technologies to embed it in other materials that make it difficult to recover (such as vitrified waste).

Some processing of nuclear materials is required to convert them into forms appropriate for many of the disposition options that have been proposed, including preparing them for disposal as waste. Processing of plutonium and uranium has historically raised environmental and public health concerns, as well as concerns about occupational health and safety. Regardless of the technology or disposition approach selected, radioactive waste will be generated and require long-term management.

Findings

Storage of plutonium and HEU from dismantled warheads will be required for one to several decades, regardless of what choices are made for ultimate disposition of these materials. DOE will present some approaches for a long-term storage facility as part of its Weapons Complex reconfiguration (in conjunction with the preparation of a Programmatic Environmental Impact Statement under the National Environmental Policy Act), and there are expected to be opportunities for public comment in that process.

Since the Administration has not made an official determination as to whether any plutonium and HEU from warheads will be declared surplus (e.g., not needed for future weapons), as yet there has been no comprehensive Federal planning process for the ultimate disposition or management of surplus materials. Discussions in and out of Government of plutonium disposition reveal little support for the use of surplus U.S. plutonium from warheads as fuel for U.S. commercial reactors. Some factors contributing to the lack of enthusiasm for this option are concerns of U.S. utilities about regulatory, public acceptance, and economic issues, as well as the fact that the United States has in the past discouraged commercial plutonium use because of proliferation concerns. DOE and certain private firms have expressed interest in the construction of special plutonium-fueled reactors at Federal sites to eliminate portions of weapons plutonium while also generating electricity.

Decisions about the fate of plutonium from U.S. weapons could influence similar decisions in Russia and other nations that may be planning to use plutonium in reactors. To reduce the world stockpile of plutonium that is readily available for weapons, actions need to be taken to discourage future production and to facilitate controlling the existing materials and making them unusable for weapons.

With respect to HEU from U.S. warheads, it is unlikely that this material will ever be considered waste. Technology is available to convert it for use as reactor fuel. However, current plans for introducing uranium extracted from Russian warheads into commercial U.S. power plants will probably precede any similar program for U.S. material. Thus, storage of HEU for several decades is a likely outcome, and safe, secure means for long-term storage must be planned.

Policy Issues and Initiatives

Eventually, the United States will have to decide what it ultimately wants to do with the stored plutonium and highly enriched uranium from its dismantled warheads. If none of it is declared surplus, presumably the plan would call for storage for an indefinite period or until it is needed for weapons. If some of the nuclear material from warheads is declared surplus, possible disposition options would include storing it indefinitely, converting it for use in existing or future reactors, or disposing of it as waste (not likely for uranium).

NUCLEAR MATERIALS STORAGE

Regardless of the ultimate disposition of plutonium and HEU from warheads, safe storage of these materials for several decades will have to be planned as soon as possible. There are many controversial and difficult issues that will take much time and effort to resolve. These include finding the most effective way to ensure safe and secure long-term storage of these materials, determining how such facilities should be regulated, and considering whether and how storage facilities can be made amenable to any bilateral or international inspections that may be agreed to in the future. Gaining public acceptance for the location of any new or modified facilities will be difficult. Because some of the issues that need to be addressed are not within the purview or expertise of DOE, it may not be desirable to confine the planning process to DOE. A broader planning process involving government agencies in addition to DOE could help identify, anticipate, and resolve key issues.

One way to provide such a process is for Congress or the President to establish an interagency task force to make recommendations about the best way to achieve safe and secure storage. The task force can also examine the feasibility and consequences of storing surplus plutonium and highly enriched uranium separately from materials reserved for stockpile requirements, and determine what type of arrangement would facilitate potential bilateral or international inspections. Also, because settling upon a suitable and acceptable location for nuclear materials storage will be a problem, the task force should consult with the public and attempt to develop a facility siting process that is agreeable to the potentially affected communities.

NUCLEAR MATERIALS DISPOSITION

In the longer term, a process will be needed to determine the ultimate disposition of surplus plutonium and HEU from warheads. So far, discussions of options have been carried on largely by technical experts and there is no consensus about most matters.

National policy on these issues is just beginning to be discussed, and the criteria against which options can be evaluated are only beginning to be considered. To help determine how nuclear materials are to be dealt with over the long term, a means should be developed to provide the President and Congress with a comprehensive basis for making the policy decisions necessary before long-term disposition of U.S. nuclear materials can begin. A preliminary step might be to obtain a broad range of governmental and nongovernmental views about what national policies, and the key criteria for evaluating them, should be. One mechanism for doing this is for the President or Congress to create a national commission that would evaluate the technical, institutional, and economic issues, and recommend goals, policies, and programs relevant to the ultimate disposition of nuclear materials from warheads.

A NEW MATERIALS MANAGEMENT ORGANIZATION

Whatever the outcome of decisions about storage and ultimate disposition of surplus plutonium and HEU, the present organization charged with this responsibility (DOE's Office of Defense Programs (DP)) may not be well suited to carry out the new nonmilitary mission of managing materials from warheads. Historically, its activities have been subject to minimal regulation, its operations have been conducted in secret, and it has not sought or welcomed public involvement or been concerned about the international implications of its actions. Its priorities continue to be maintaining the warhead stockpile.

In contrast, the mission of storing and dealing with surplus materials is essentially civilian in nature, and potentially subject to extensive domestic regulation and to scrutiny by the international community. It may be best to have an organization that is structured from the start to do this job in a way that gives priority to ensuring safety and protecting human health and the environment, operating in an open manner, involving the public more effectively, responding to public concerns, and being constantly aware of the international implications of its activities. Such an organization could be created within DOE. Alternatively, Congress could create an organization outside DOE (perhaps in some existing agency) to carry out activities related to the disposition of surplus nuclear materials from dismantled warheads.

INFORMATION ACCESS

Status

The institutional framework for making decisions about nuclear warhead dismantlement and materials disposition is essentially the same as it was throughout the Cold War. The decisionmaking structure has historically been characterized by lack of regulation or outside oversight, restricted public access to information, and little if any public involvement. Current restrictions on access to information relevant to nuclear warhead dismantlement and materials disposition are based on legislative requirements generally intended to protect national security during the Cold War.

Findings

The executive branch has undertaken some reviews of various Federal agency procedures related to classification and declassification of information, but those efforts are typically slow and may not address public concerns about the lack of information access in warhead dismantlement and materials management matters, particularly with respect to environmental, health, and safety issues.

Many of the restrictions on information enacted to meet the Cold War situation may no longer be necessary to preserve national security, although certain types of information about warhead design and manufacture must still be withheld because of potential terrorist activities and other security concerns. However, a great deal of information relevant to warhead dismantlement and materials management could be made more accessible, particularly data having to do with the environment, health, and safety.

Policy Issues and Initiatives

In light of the increased authority of the States and of the public in activities at the Nuclear Weapons Complex, DOE will have to plan and conduct its dismantlement and materials management activities in a more open manner that will permit more public involvement. To facilitate public access to relevant information, legislative and administrative restrictions on information access should be evaluated to determine what changes are needed to suit the new circumstances of the post-Cold War era and enhance public involvement. Although the Administration is reviewing some of these matters, more attention could be devoted to efforts to revise current standards and procedures for access to information specifically related to warhead dismantlement and materials management. Also, Congress could review the Atomic Energy Act and other pertinent laws, or request that the Administration conduct such a review, and recommend changes to facilitate public access to appropriate data relevant to nuclear warhead dismantlement and materials management.

COOPERATION WITH RUSSIA

Status

Russia has announced plans to retire and dismantle a substantial portion of its nuclear weapons stockpile over the next decade or more. The United States has pledged several hundred million dollars for technical assistance in this connection, but only a small portion has been spent.

The Russians have indicated that a lack of storage for their nuclear materials, especially plutonium, is impeding their ability to dismantle warheads. After a series of discussions, the United States and Russia have agreed that Russia will design its own storage facility for special nuclear materials from warheads, with design assistance from the United States provided through the U.S. Army Corps of Engineers. The Russians plan to begin site preparation for this storage facility within a year, but many political, technical, and financial obstacles could hinder its successful completion in the near term.

U.S.-Russian agreements have also been reached on U.S. provision of, or assistance with, specific items such as armored blankets, warhead storage containers, emergency response systems, and secure rail cars to enhance the safe transport of weapons. However, these efforts have not had any significant effect yet on Russian warhead dismantlement—an objective that requires continuous emphasis at the highest levels of U.S. Government and by the several agencies designated to conduct the Russian assistance program.

With respect to the HEU from Russian weapons, the United States and Russia entered into an agreement in February 1993 (subject to terms not yet finalized) whereby 500 metric tons of the material would be converted to low-enriched uranium (LEU) in Russian facilities and then purchased by the United States. At least 10 metric tons would have to be converted in each of the first five years and 30 metric tons in each of the following years (for a total of 20 years). A final purchase agreement has yet to be executed, however.

An implementing agreement would specify price, certain conditions, and a method of sharing proceeds among other former Soviet republics. The contract is intended to provide for participation by both the U.S. private sector and Russian enterprises; it is also intended to establish "transparency measures" for materials control and accounting.

Findings

While the United States views expeditious Russian warhead dismantlement and materials disposition as vital to its national security, Russia's agenda is dominated by economic and political issues that could diminish the priority given to dismantlement. U.S. efforts to assist Russia with nuclear warhead dismantlement have initiated an important cooperative process but have not yet had a significant effect on the Russian dismantlement program. And they have not been carried out in a manner that addresses the broader issues of mutual goals and interests in stockpile reduction and materials management.

The U.S. purchase of Russian HEU from warheads is nearing final agreement and will contribute to a reduction of the materials available for new nuclear weapons there. However, it will be decades before large portions of the total Russian inventory of this material are converted and transferred to the United States, and therefore a significant risk of diversion will remain. The United States appears to have entered into this agreement without a fully articulated analysis of what further steps might be taken to improve the management and control of warhead materials to prevent their diversion.

The United States has not verified specific warhead dismantlement activities and accomplishments in Russia, and has no direct cooperative process for developing accurate information about Russian dismantlement status and capabilities. Further, the United States has not established a policy or approach to mutual dismantlement verification, warhead materials storage, or other materials management and control activities (including possible future production of warhead materials).

Efforts to integrate U.S. warhead dismantlement progress and plans with programs to assist Russia have not received adequate attention. There is also little linkage between Russian economic, environmental, or social needs and U.S. programs to assist or encourage Russian dismantlement and other related activities.

Policy Issues and Initiatives

The United States needs a plan for helping Russia's dismantlement and materials management process to proceed safely and without allowing warheads or warhead materials to get into the wrong hands. An important aspect of the plan is to increase coordination between the agencies responsible for U.S. materials management and disposition programs and those responsible for U.S. policy toward Russia. This is important because the United States must develop policies that utilize U.S. experience in its programs to assist Russia.

Because many problems and needs in Russia are unrelated to dismantlement, it is also important at this time to strengthen the link between U.S. assistance in nuclear materials disposition programs and other programs in which assistance is desired by Russia. It would help if there were cooperative efforts between the two nations in these matters.

To carry cooperative efforts further, an arrangement could be worked out with Russia whereby the United States would fund a 2-year joint study of materials disposition scenarios to be conducted by a U.S.-Russian multidisciplinary team based in Russia. To help ensure that dismantlement and materials disposition are proceeding safely and securely, the United States could also develop and negotiate with Russia an initiative for mutual disclosure of the amounts of weapons plutonium and highly enriched uranium possessed by each country.

An important issue is whether any storage or processing facilities used in connection with warhead dismantlement and materials management should be subject to international monitoring, inspections, or even control. In that regard, it remains to be seen whether the United States can realistically expect to verify, either directly or through international agencies, Russia's compliance with a specified rate of dismantlement—and its controlled storage of special nuclear material—without some reciprocal interest by Russia in verifying U.S. progress along the same lines. A high-level governmental process is needed to consider and address means to achieve reciprocal arrangements to verify the amounts and monitor the status of these materials in the future.

CONCLUSION

Reducing the nuclear weapons stockpile will not be simple, painless, or inexpensive. Although the work of retiring and disassembling weapons that are outdated or no longer needed in the stockpile is under way, the next critical steps in the process are uncertain because no national policy exists to guide future dismantlement and materials management activities in the United States. In addition, the United States has not developed an effective strategy for encouraging and assisting Russia in its efforts to safely dismantle its warheads, and to safely and securely manage the materials from them.

It is important that warhead dismantlement and materials management be conducted successfully both here and abroad. Failure to do the job right in the United States could create risks of accidents, dangers to workers, and harm to the environment and populations. In Russia, all of these risks exist, but there are also risks that the weapons or materials could be diverted and fall into the wrong hands.

Yet, the existing approach by the United States to both U.S. and Russian dismantlement is insufficient. As yet, the Nation has no coordinated, comprehensive policy on this subject and there has been little informed public debate on the establishment of national goals.

The prospects for successfully carrying out dismantlement and materials management activities in the future—and perhaps assisting Russia in similar efforts—can be improved if leadership is provided now at the highest levels of government. Policy guidance will be needed from these levels. To provide such guidance, the Federal Government will first have to articulate a national policy on dismantlement—a policy that sets the objectives and rationale for permanent stockpile reduction.

In sum, the challenge ahead requires planning and decisions in the near term if it is to be successful in the long term. The process deserves consistent and enduring talent, dedication, and resources, as well as astute management.

Analysis of Warhead Retirement Programs and Plans 2

Recent treaty agreements and announcements by the United States and Russia to substantially reduce the nuclear weapons arsenals of both nations (see box 2-A) present a unique opportunity, as well as a technical and political challenge, to the Nation. The opportunity is to remove many thousands of these weapons from current arsenals. The challenge is to devise feasible and practical methods of rendering existing weapons benign and to formulate reasonable plans and make decisions regarding the short- and long-range goals of this effort (17,20).

Decisions to retire and dismantle large numbers of nuclear weapons have provided the incentive to define and initiate major Federal programs. Both the Departments of Energy (DOE) and Defense (DOD) have established offices and task forces to plan and implement various steps in dismantlement.[1] Thousands of weapons have been taken out of the active stockpile; a few thousand warheads have been retired and disassembled; and the remaining warheads from retired weapons will be disassembled over the next decade. Table 2-1 shows the major types of weapons in the current U.S. arsenal.

Even though world leaders have made these announcements about the elimination of nuclear weapons and declared that the "Cold War is over," the recent Strategic Arms Reduction Treaty (START) and its successor (START II) have not yet been ratified

[1] For example, the Executive Management Team for Dismantlement under Defense Programs; the Plutonium (and highly enriched uranium) Strategy Task Forces under DOE (Defense Programs and Nuclear Energy Program); Midnight Sun—the Defense Nuclear Agency (DNA) or the Dismantlement Planning Effort; the Prioritization Working Group of DNA and DOE; the Nuclear Weapons Safety Committee, established by DOE.

Point

"Although major emphasis within the nuclear community has shifted from development to retirement and disposal of weapons, I do not view it as a major change in mission, but more of a shift in national emphasis."

Pentagon reviewer of OTA report

Counterpoint

"This is no time for complacency. The arms race may be over, but it is still too much business as usual for the nuclear weapons industry. The government has yet to indicate that it can handle conversion to peacetime work."

A citizens' guide to the future of the Nuclear Weapons Complex

Box 2-A—Proposed Reductions in U.S. and Russian Nuclear Arsenals

In just the past 2 years, political leaders in the United States and the former Soviet Union have made remarkable pledges and set an agenda for denuclearization that, if completed, will eliminate a large part of the nuclear weaponry built during the Cold War. Some of the key pledges follow (1):

- In the fall of 1991, Presidents Bush and Gorbachev pledged to withdraw almost all tactical nuclear weapons from forward deployment and destroy most of them.
- As the Soviet Union dissolved toward the end of 1991, leaders of the newly independent states, where nuclear weapons were based, pledged to join the international community as nonnuclear weapons nations and to remove such weapons from their territories.
- In May 1992, the current nuclear nations pledged to ratify the Strategic Arms Reduction Treaty (START) that calls for sharply reducing the strategic arsenals of the United States and the former Soviet Union.
- In June 1992, Presidents Bush and Yeltsin agreed to the outline of a START II treaty that would eliminate all but 3,000 to 3,500 nuclear weapons deployed (deliverable) in each of their remaining strategic arsenals.
- Presidents Bush and Yeltsin signed the START II treaty in Moscow in January 1993.

SOURCE: Office of Technology Assessment, 1993.

and no international agreements exist on the subject of warhead dismantlement. However, there is substantial agreement among interested parties within and outside of government that nuclear weapons retirement and dismantlement in both the United States and the former Soviet Union are important steps in promoting national security and world peace.

This chapter discusses the programs and plans now in place to retire nuclear weapons and return the warheads to DOE for dismantlement. The programs are mainly the responsibility of the Department of Defense and the Department of Energy, which is the Nation's nuclear warhead design, construction, and testing agency.

The current process of dismantling nuclear weapons begins with an action by DOD to retire a weapon from the active stockpile and, when appropriate, return it to a military base within the continental United States.[2] By making such a retirement decision, DOD takes an action that eventually leads to return of the weapon to DOE, which originally built the warhead and will retain custody until it has been dismantled and its components have been disposed of in a manner determined by DOE. The individual services (Army, Navy, Air Force) have physical custody of these weapons until they are transferred to DOE (5).

There is a long-standing administrative process for the management, handling, and control of nuclear weapons within the Federal agencies having such responsibilities, principally DOE and DOD. These agencies are now beginning to modify their procedures for the new, post-Cold War mission of dismantlement and disposal. Figure 2-1 illustrates this administrative process. It begins with preparing a Nuclear Weapons Stockpile Memorandum—a document that is revised and updated each year and contains a 5-year projection of the U.S. stockpile. The three-member Nuclear Weapons Council (the Assistant Secretary of Defense for Logistics, the Vice-Chairman of the Joint Chiefs of Staff, and the Assistant Secretary of Energy for Defense Programs) considers and submits the memo to the President through the Secretaries of Defense and

[2] The term "retirement" as it relates to nuclear weapons within DOD is an administrative decision to change the weapon's charge code. This change in charge code is from the active stockpile to either the retired stockpile or one of the reserve stockpiles. DOD has the responsibility of removing the warhead from the delivery system and disposing of that delivery system. Activities associated with launcher disposal are not covered in this study but are, nonetheless, an important part of the total scheme.

Table 2-1—Major Nuclear Weapons Types in the Current U.S. Arsenal

Designation	System/common name	Service	Primary uses
W48	155-mm AFAP	A	Surface to surface
B53	Strategic bomb	AF	Air to surface
W56	MINUTEMAN II ICBM	AF	Surface to surface
B57	Bomb/depth bomb	N, AF	Air to surface/subsurface
B61	Tactical/strategic bombs	N, AF	Air to surface
W62	MINUTEMAN III ICBM	AF	Surface to surface
W68	POSEIDON C3 SLBM	N	Underwater to surface
W69	SRAM	AF	Air to surface
W70	LANCE	A	Surface to surface
W71	SPARTAN	A	Surface to air
W76	TRIDENT 1 C4 SLBM	N	Underwater to surface
W78	MINUTEMAN III ICBM	AF	Surface to surface
W79	8-inch AFAP	A	Surface to surface
W80-0	TOMAHAWK (TLAM N)	N	Underwater to surface/ surface to surface
W80-1	ALCM	AF	Air to surface
B83	Strategic bomb	AF	Air to surface
W84	GLCM	AF	Surface to surface
W87	PEACEKEEPER ICBM	AF	Surface to surface
W88	TRIDENT II D5 SLBM	N	Underwater to surface

NOTE: This is an official unclassified list of weapons types, with the older ones at the top and the newer ones at the bottom. It should be noted that all Army (A) weapons have been retired and the Marine Corps' nuclear mission has been deleted. Both the Navy (N) and the Air Force (AF) are reducing the total numbers of weapons in their stockpiles. Total stockpile numbers are classified.

SOURCE: Defense Nuclear Agency, 1993.

Energy and the National Security Council. When the President approves, it becomes the Nuclear Weapons Stockpile Plan (NWSP) and is executed by the Assistant Secretary of Defense (Atomic Energy).

The Office of the Assistant to the Secretary of Defense (Atomic Energy) (ATSD(AE)) performs a coordination function within the Office of the Secretary of Defense before the memorandum is submitted to the Secretary of Defense and Secretary of Energy for approval/transmission to the President. DOE also coordinates internally before the Secretary of Energy signs it. Once signed by the President, the document is addressed and delivered to the Secretary of Defense, Secretary of Energy, and Chairman of the Joint Chiefs of Staff. ATSD(AE) gets the copy sent to the Secretary of Defense and sends it to the services. It forms the basis of the allocation plan that is sent to the services and the Defense Nuclear Agency (DNA) for tracking (6).

As shown in figure 2-1, the NWSP generates other procedures and actions through DNA, the various military services, and DOE. The process eventually results in directives to retire specific weapons, return the warheads from DOD to DOE, and begin dismantlement under a schedule that accommodates the capabilities and constraints of all the parties involved (5).

Both DOD and DOE have embarked on planning efforts to ensure that the process will proceed safely and effectively. Plans continue to be

Figure 2-1—Planning and Coordination Process for Nuclear Weapons Retirement

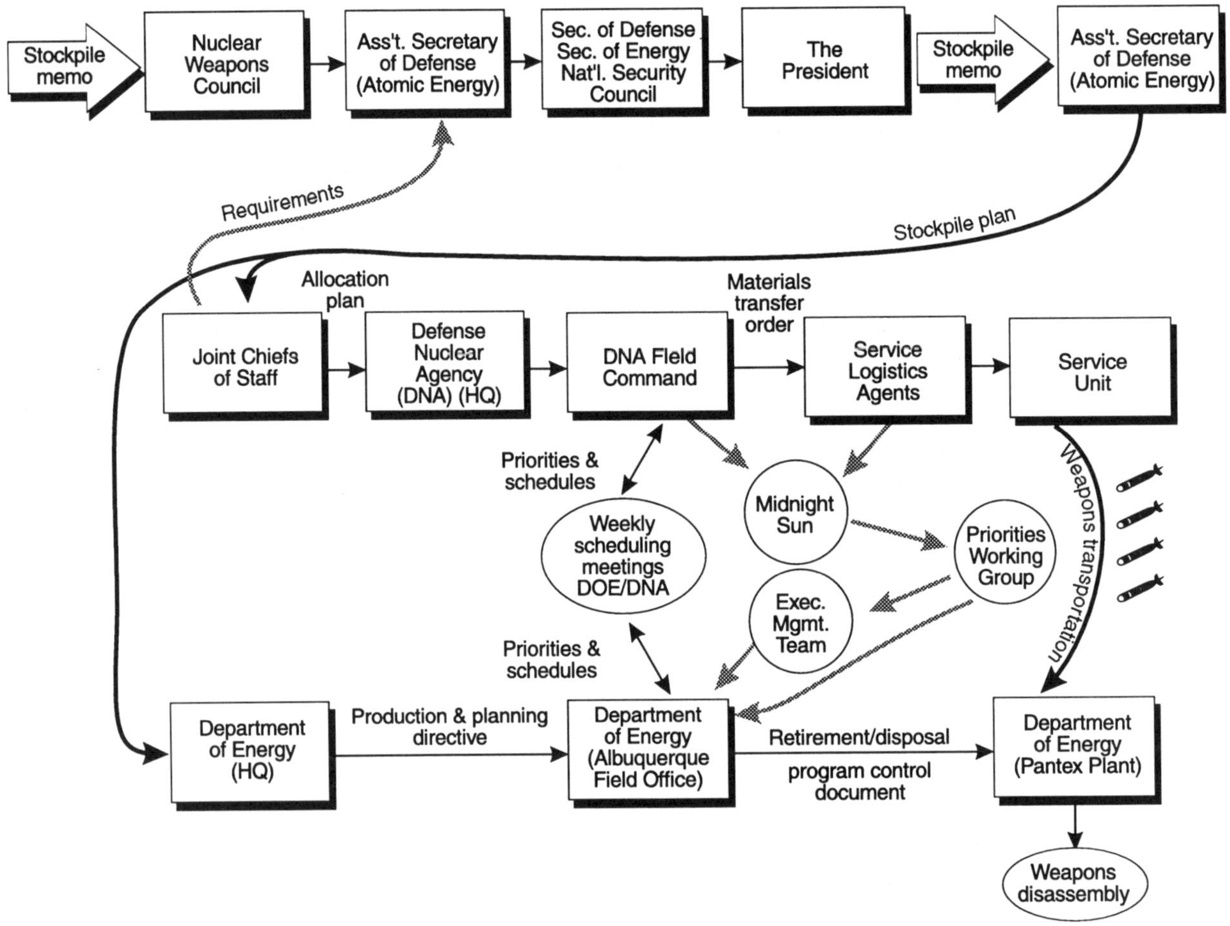

Midnight Sun: the name of the DNA working group charged with planning and monitoring weapons retirement and return
DOE Executive Management Team: DOE's dismantlement planning group
Priorities Working Group: the joint DOE/DNA group charged with planning and logistics

SOURCE: U.S. Department of Defense, Defense Nuclear Agency.

updated as variables such as base closures have their effect. Actual weapons retirements are now under way at various military bases, warheads are being transported to DOE, and dismantlement operations at DOE facilities are following suit. It is not clear, however, when the new planning and review activities will be incorporated into such operations, or how carefully the new mission will be defined so as to result in the needed management and institutional changes. Box 2-B illustrates some of the challenges ahead.

In some ways, the work required to retire and dismantle weapons is similar to that of producing weapons and maintaining a stockpile. Over the past four or five decades, some classes of nuclear weapons have been retired and others have taken

Box 2-B—The Challenge of Stockpile Reduction

Pledges to rid the world of massive numbers of nuclear weapons have been made with great fanfare and hopes for lasting peace. The challenge of turning these pledges into deeds and accomplishing the goals, as viewed in mid-1993, is formidable (1):

- Several tens of thousands of nuclear warheads are located at hundreds of sites worldwide. These warheads have massive explosive power, and their continued existence—especially in politically unstable areas of the world—poses serious dangers. The threat is not only from certain governmental or individual terrorists who may be able to detonate a weapon, but also from either accidental or intentional dispersal of radioactive materials.
- If not carefully controlled, the stockpiles of highly enriched uranium and plutonium produced over the past 50 years could be diverted to terrorist groups that may have the ability to make crude but dangerous bombs.
- Radioactive materials from weapons can pose substantial long-term threats to human health and the environment if not safely contained and managed, and no direct methods are currently available for destroying them.
- Worldwide, more than a million people work in laboratories, facilities, and factories engaged in nuclear materials or weapons production and maintenance. These workers and their families face an uncertain future, a changing mission, and a threat to their livelihood and position in society.
- Thousands of nuclear-weapons-related facilities and huge expanses of the environment are contaminated with radioactive and toxic waste from 50 years of plutonium and tritium production. This poses serious environmental and public health threats that are just beginning to be addressed. As nations prepare to dismantle warheads and dispose of materials from them, they will need to focus serious attention and resources on requirements for human health and safety, which will involve both a change of attitude and an assumption of hitherto neglected environmental responsibilities.

SOURCE: Office of Technology Assessment, 1993.

their place. Representative weapons of all types have been dismantled for purposes of quality assurance and reliability testing. In some cases, conventional weapons systems have replaced retired systems. Parts and materials have been recycled whenever advantageous, and significant efforts have been devoted to careful disassembly and accounting, as well as the handling of sensitive parts and materials. Accurate accounting for fissile materials has been basic to weapons materials security.

Both DOD and DOE claim that because weapons have always been dismantled as part of their replacement or maintenance—and because materials from these weapons have always had to be recycled or disposed of in some way—there is nothing very different about the present mission. Thus dismantlement is considered by many to be merely a change in emphasis and an adjunct to the technical operations (4,7,8). Yet in many ways, the work represents a significant change of focus from the past missions of these agencies.

The new mission for these agencies is to significantly reduce the overall stockpile and manage materials that are no longer part of national defense. This mission offers new challenges and requires different technical and management skills. It also brings a need for planning that rests on an entirely different premise from the weapons production mission. Planning is necessary now to ensure that more nuclear materials are safely stored and controlled outside the military system, that goals for stockpile reduction and disposition of materials are met in the United States and elsewhere, and that past mistakes leading to environmental degradation or threats to public health are not repeated (15).

DOD: WEAPONS RETURN AND PREPARATIONS FOR DISMANTLEMENT

The Department of Defense is responsible for the separation of warheads from their delivery systems, such as missiles, for which it retains responsibility. DOD is currently implementing substantial reductions of the U.S. nuclear arsenal. However, neither the international agreements concerning the reduction of deployed weapons nor the START and Intermediate Range Nuclear Forces (INF) treaties specify that warheads be dismantled—they merely call for removing warheads from delivery systems (e.g., missiles) and, in some cases, destroying the delivery systems. Certain experts, in fact, argue that dismantlement goals should not fall within the terms of these agreements (7).[3]

DOD has stated in the past that if the United States had to "destroy" existing warheads that are being "retired" under current treaties and agreements, it would have to produce new warheads to maintain its reduced inventory. Yet the U.S. Nuclear Weapons Complex currently has little, if any, operational capability to produce new warheads. There is also concern on the part of some military planners that future weapons readiness capabilities, and even U.S. military supremacy, could suffer if too many nuclear warheads are destroyed before future world threats have been carefully evaluated (7).

The uncertainty about decisions to retire and dismantle nuclear weapons—as well as vagueness in the definitions of these and other terms—means that long-range plans, goals, or capabilities cannot be accurately determined or evaluated at this time. The near-term technical questions are, therefore, whether the retirement and dismantlement operations now under way are being carried out safely and effectively, and whether the United States will be able to safely and adequately prepare for the next step in the process—the long-term disposition of nuclear materials from warheads.

Important policy questions are:

- whether informed public debate will be brought to bear on major retirement and dismantlement decisions;
- whether it will be necessary to develop new policies to direct and coordinate this work;
- whether the retirement and dismantlement process now under way will proceed efficiently and effectively, with adequate attention to health, safety, and environmental protection; and
- whether the effort will sufficiently advance the stated national security goal of international reduction of nuclear armaments.

Number of Weapons To Be Retired or Dismantled

No official government list of U.S. nuclear weapons to be retired or dismantled is publicly available. It is generally agreed, however, that the stockpile will be reduced substantially, but the precise number of each type of weapon (retained or dismantled) is reflected in the Nuclear Weapons Stockpile Plan, prepared by the joint DOD-DOE Nuclear Weapons Council and approved by the President. That document is classified, and the plan is continually modified to reflect the most current U.S. policies and international understandings (18).[4]

As a beginning, DOE has developed a general policy for dismantlement. In its policy, DOE has stated that it will dismantle all retired warheads turned over to it by the Department of Defense.

[3] DOD claims it needs the flexibility to reuse some warheads from retired weapons if it so chooses; that the greatest cost in weapons is delivery vehicles, not warheads (which are easy to reassemble); and that elimination of warheads is of no use unless constraints are also placed on fissile material production.

[4] The U.S.-Russian agreement, signed in June 1992 and codified by the START II treaty signed in January 1993 by Presidents Bush and Yeltsin, reduces the size of each nation's nuclear arsenal to 3,000 to 3,500 deployed (deliverable) strategic warheads by 2003. The reductions called for are the most sweeping in U.S.-Russian bilateral arms control history. They represent a major reduction in present arsenals.

U.S. DEPARTMENT OF ENERGY

Retired nuclear weapons received at the Department of Energy Pantex Plant.

Dismantlement will consist of disassembling each warhead, removing and storing the uranium and plutonium components, and disposing of the remaining parts in accordance with State and Federal regulations. Most of the weapons retired by DOD are now being shipped (or will be shipped) to DOE's Pantex Plant near Amarillo, Texas for dismantlement (one weapons type was recently dismantled at the Y-12 Plant in Oak Ridge, Tennessee). These shipments are planned and controlled jointly by DOD and DOE to ensure safe and effective weapons dismantlement. DOE has determined that its current safe maximum rate is approximately 2,000 weapons per year. According to DOE, this rate was selected to ensure that dismantlement could be carried out in an orderly, safe, and environmentally sound manner. DOE has also said that this rate will allow the United States to complete the dismantlement of anticipated planned retirements by the end of this decade (4).

On further investigation, the above rate appears to be optimistic over the near term. Pantex managers have indicated that the current targets are 1,700 per year in FY 1993 and 2000 in FY 1994. The FY 1992 rate was about 1,300 at Pantex (see table 2-2 and figure 2-2). At the current FY 1993 Pantex dismantlement rate, the year-end total would be about 1,430 warheads.

Current U.S. warhead dismantlement plans are governed by reductions in the nuclear stockpile planned by the Nuclear Weapons Council. Some of these plans also take into account the general goals of the 1988 INF treaty, the 1991 announcement of unilateral withdrawal of tactical weapons by President Bush, and both the 1991 START treaty and the START II treaty signed by Presidents Bush and Yeltsin in January 1993 (see table 2-3). Taken together, these initiatives, if ratified and fully implemented, could lead to the eventual dismantlement of more than 10,000 U.S. warheads by early in the next century (16). In fact, START II sets a limit of no more than 3,500 deployed (deliverable) strategic nuclear warheads each for the United States and Russia by the year 2003. The actual number of warheads to be returned to DOE for dismantlement is specified in the annual Nuclear Weapons Stockpile Memorandum. This classified memorandum is submitted to the President for approval each year on September 30.[5] The most recent two submissions have included an attachment listing the specific weapons to be retired. The President does not directly approve or disapprove this list, but could consider it in review of overall stockpile strategy. If the above treaties are fully implemented, the results should be reflected in the annual stockpile memorandum, which also contains specific numbers of nuclear weapons projected to be in the active stockpile, the reserve stockpile, and retired.

At present, the return of retired nuclear warheads to DOE is well under way. All retired weapons that were to be returned to the United States under former President Bush's initiative had been returned to continental U.S. bases by the end of 1992. Since 1990, DOD has been shipping retired warheads from military bases to the Pantex Plant for dismantlement. Almost 4,000 retired warheads were dismantled at Pantex from 1990

[5] See classified Annex to this report for a discussion of current weapons stockpile amounts and future plans.

Table 2-2—DOE's Pantex Plant: Nuclear Warheads Disassembly History, 1990-92

Fiscal year	Numbers retired and disassembled
1980	535
1981	1,416
1982	1,360
1983	960
1984	860
1985	927
1986	574
1987	1,068
1988	510
1989	1,134
1990	1,056
1991	1,546
1992	1,274
Total	**13,223**

NOTE: In addition to these weapons that were disassembled subsequent to retirement, additional warheads were disassembled for quality assurance and reliability testing each year—some in this category were disposed of and others were reassembled and returned to the stockpile. The number in this category ranged from less than 100 to about 400 per year.

SOURCE: U.S. Department of Energy.

through mid-1993 (table 2-2). If the current retirement plans are followed, at least 10,000 warheads should be returned for dismantlement over the next decade.

The total numbers of warheads in the active and reserve stockpiles currently, and in the past, are classified. The number of each type and design of warhead that has been retired and dismantled is classified. Annual warhead dismantlement rates at Pantex are available but not in a form that would allow one to calculate back to actual stockpile numbers. Even so, it is clear that the challenge of effectively and safely managing a return, dismantlement, and materials disposition program of the magnitude resulting from the above stockpile reduction goals is daunting.

This study by the Office of Technology Assessment does not attempt to determine dismantlement quantities and rates beyond the very general estimates given above. If all of these data remain classified, an important question is how public policy will be established for the future storage, control, and safe management of warheads and of the materials from their dismantlement. If there is to be a public discussion, however, useful order-of-magnitude information about the quantities, rates, and storage or processing requirements of weapons to be dismantled may need to be made public and official.

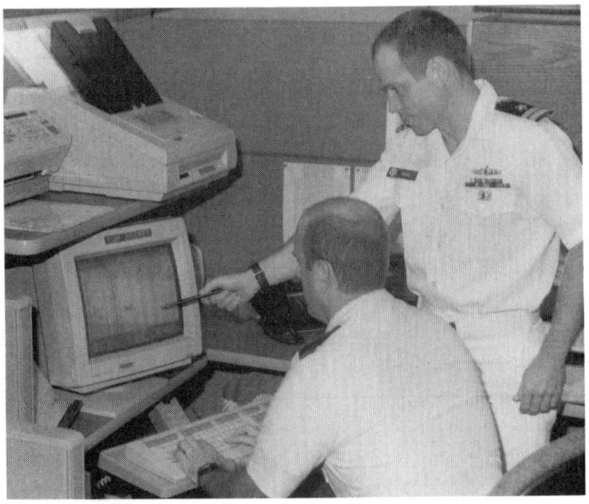

U.S. DEFENSE NUCLEAR AGENCY

DOD planners evaluating schedules for weapons returned to DOE.

Current DOD Plans and Programs

Because the information is classified, the Defense Department will not divulge the location, storage, or transportation routes of nuclear weapons. However, the movement of weapons from overseas to continental U.S. bases has been completed. Further retirement of weapons will continue to take place under the guidance of stockpile reduction policies. The individual services with custody of the weapons have storage and maintenance facilities for all weapons whether the weapons are active, active reserve, inactive reserve, or retired awaiting DOE pickup prior to dismantlement. Because this information is classified, the storage capacity relative to the number

Figure 2-2—Pantex Warhead Dismantlement FY 1992

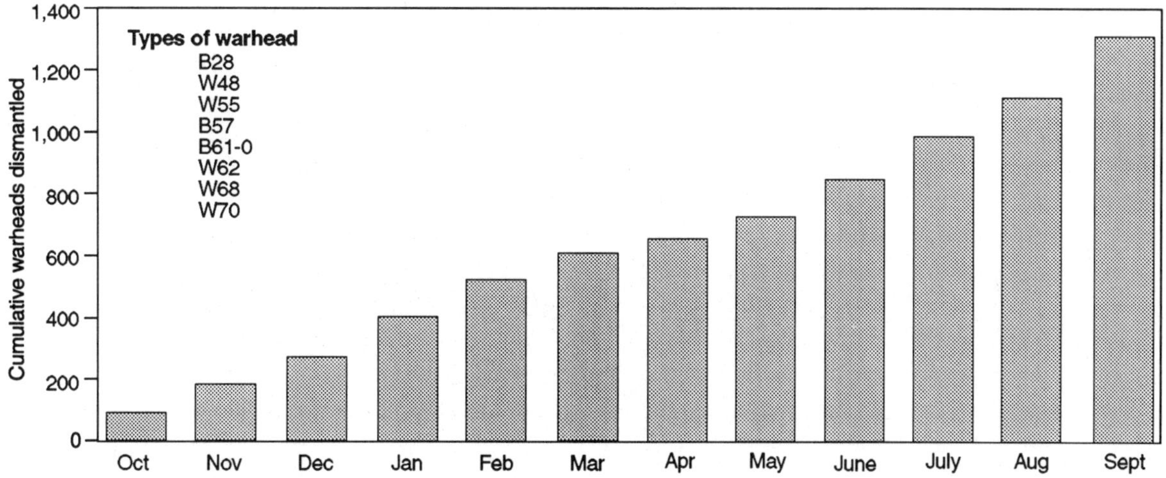

SOURCE: U.S. Department of Energy.

of warheads to be retired cannot be revealed. Currently, the armed services believe they can store projected warhead retirements, given current base closures and the DOE dismantlement capability. However, any delays in dismantlement or changes in base closure or base operation consolidation plans could impact DOD's capability to store its warheads. The number of active DOD nuclear capable storage sites is gradually decreasing, primarily through the services' effort to consolidate operations. The Army is heading toward complete elimination of its nuclear arsenal and the other services are closing certain of their storage sites (5).

Figure 2-3 shows the process of weapons retirement in DOD and dismantlement in DOE, and indicates some of the major issues affecting the key steps. For DOD, the questions of transportation, safety, and security, as well as pressures to move weapons because of facility closures, appear to be driving many of the logistical decisions.

NUCLEAR WEAPONS SAFETY ISSUES

Nuclear weapons have been handled by the U.S. military services over the past four decades with great attention to safety of operations. The DOD process of retiring large numbers of weapons and returning warheads for dismantlement, however, brings a responsibility to review the suitability and comprehensiveness of safety practices.

Safety questions have long had high priority among agencies with nuclear weapons responsibilities. The Nuclear Weapons System Safety Group maintains a safety report on each weapon. The Defense Nuclear Agency reviews safety reports from the Nuclear Weapons Systems Safety Group and performs safety analyses. Many types of safety analyses are performed regularly.

DOD is responsible for returning nuclear weapons located outside the continental United States. Weapons are returned via air or sea transport to bases at designated locations within the continental United States. The intact warhead delivery system is considered rugged enough to be its own shipping container in the case of warheads placed in missile reentry vehicles or in air-dropped bombs. Other types of warheads are placed in special containers, when appropriate, prior to shipment back to the United States from Europe. Some weapons, although rugged enough to constitute their own shipping containers, require other containers, for efficient stacking.

Table 2-3—START-II Deployed (Deliverable) Warhead Limits
(for each country—the United States and Russia)

Warhead type	Limits	
	Phase I[a]	Phase II[b]
Strategic total (warheads attributed to deployed ICBMs, deployed SLBMs, and deployed heavy bombers)	3,800-4,250	3,000-3,500
MIRVed ICBMs	1,200	0
SLBMs	2,160	1,700-1,750
Heavy ICBMs	650	0

NOTE: ICBMs = intercontinental ballistic missiles. SLBMs = submarine-launched ballistic missiles. MIRV = multiple independently targetable reentry vehicle.
[a] To be completed 7 years after the treaty becomes effective.
[b] To be completed by the year 2003.

SOURCE: U.S. Arms Control and Disarmament Agency.

For the expanded retirement program just completed, DOD used existing planes and bases, with limits dictated by safety considerations. DOD and DOE are also using Safe Secure Trailers to the maximum extent for transportation within the continental United States. According to Pentagon officials, as of mid-1993, all of the planned weapons' returns from overseas to U.S. home bases had been accomplished. The movement of weapons by DOD will now focus on the consolidation of U.S. bases with a future goal of a minimum number of nuclear weapons storage sites for each service within the United States (2,3,8).

In the past, concerns about accidents with nuclear weapons caused DOD to make certain safety improvements. For example, plutonium scatter accidents, in which accidental detonation of the chemical explosives in a warhead disperses plutonium and other nuclear materials into the environment, are of particular concern during nuclear weapons transportation. Such accidents involving aircraft on alert have occurred in the past. For example, a bomber carrying nuclear weapons crashed near Palomares, Spain, in 1966 during a refueling exercise, and another crashed during takeoff in Thule, Greenland, in the late 1960s. Even though these accidental detonations of high explosives did not lead to actual nuclear explosions, they did result in widespread dispersal of the weapons' plutonium, extensive environmental contamination, and high cleanup costs (14).

The practice of airborne alert has been discontinued, and the Pentagon reports that no aircraft crashes have occurred during logistic movements of nuclear weapons. One response to dispersal accidents was to develop new types of high explosives for use in nuclear warheads. The nature of the chemical high explosives used in a particular warhead is of critical importance to the risk of plutonium scatter accidents. All nuclear warheads produced before 1979 contain an older-design, conventional chemical high explosive (HE) that can detonate under some accidental conditions, including airplane crashes or fires, causing plutonium to scatter. The most modern warhead designs utilize insensitive high explosive (IHE), which is safer since it is designed so that detonation will not occur under similar accident scenarios (9,13). Research efforts at Sandia National Laboratory have developed approaches to minimize plutonium scatter accidents. One past improvement was the development of accident-resistant air shipping containers

Figure 2-3—The Nuclear Stockpile Dismantlement Process (with major issues in parentheses)

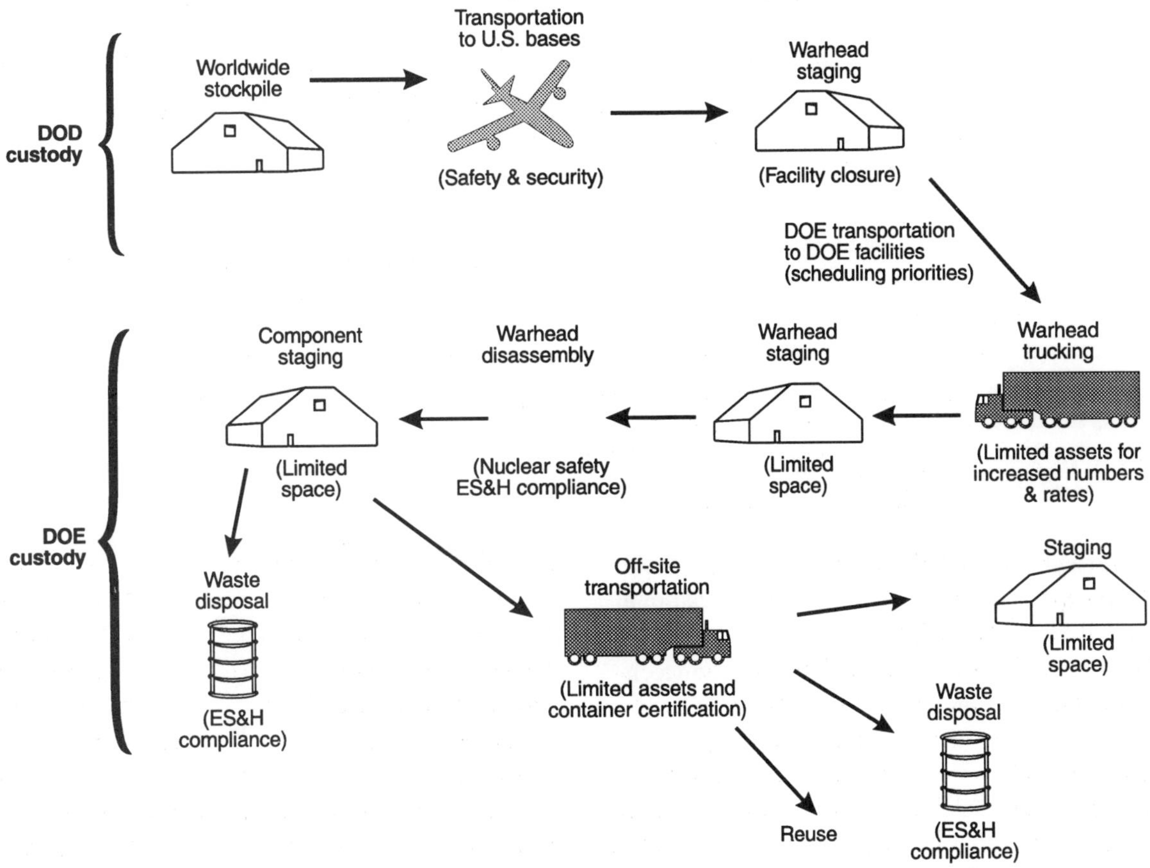

NOTE: ES&H = environment, safety and health.

SOURCE: U.S. Department of Energy.

that were designed to prevent dispersal of nuclear materials should an accident occur.

Weapons with conventional HE could be a problem under certain conditions if these accident-resistant containers are not used or other precautions are not taken (9). Even though the services have used such containers when necessary, DOD does not require the use of these special containers even if the weapons contain conventional HE.

A more recent analysis of safety issues conducted in 1990—and known as the "Drell Report" (19)—was an independent study by the Panel on Nuclear Weapons Safety requested by the House Committee on Armed Services, joined by the Senate Committee on Armed Services. It considered safety issues as part of developing future U.S. nuclear forces in the context of recent profound changes in the strategic, political, and military dimensions of international security. The report painted a picture of a weapons program that had in the past been far more concerned with production than with safety. Its organizational recommendations are now being implemented by the Defense Nuclear Agency and others. They included establishing or improving the organizations and procedures for evaluating and correcting defects, enhancing training programs, and evaluating new concerns for operations and functions in the post-Cold War world. It will be important to support these efforts in the future if safety is to remain in the forefront.

Based on the recommendations in the Drell Report, DOD and DOE have recently established two advisory committees on nuclear weapons safety. The first is the DOE-DOD System Safety Red Team Advisory Committee. The Red Team is responsible for the technical evaluation of weapons designs and procedures on preventing inadvertent detonation or plutonium dispersal, and for reviewing the safety of warhead and subsystem designs in all credible environments as well as the documentation related to such subsystems (5). The second committee is the Joint Advisory Committee on Nuclear Weapons Surety. This committee has responsibility for advising on inadvertent detonation and plutonium dispersal, and for making recommendations on national policies and procedures to ensure safe handling, stockpiling, maintenance, and risk reduction technologies for nuclear weapons (see box 2-C).

Transportation Safety

The responsible services within DOD take great care to ensure the safety of transport when weapons or warheads are moved between military bases or depots, or between staging sites and destinations within the United States or overseas. Because larger numbers of weapons are being retired and weapons are being moved more frequently, transportation safety has received even more attention.

Over the last few years, some comprehensive assessments have been accomplished and new directives have been proposed to modify and improve transportation safety. Much work has been done to analyze what changes could be effective, but most have not yet been fully approved and put in place. Also, significant controversy exists about the merits of certain proposals, whereas others have generally been accepted but are being implemented slowly.

U.S. DEPARTMENT OF ENERGY

A specially equipped DOE transport vehicle used to ship retired weapons to Pantex as well as to ship components to other sites.

As of mid-1993, a major study (begun in 1988 and completed at the end of 1992) still awaits approval of the Nuclear Weapons Council (10). This joint DOE-DOD Study on the Logistic Transportation of Nuclear Weapons represents the first rigorous and formal probabilistic risk assessment ever done on the subject. It incorporates an extensive database on accidents and their probabilities; assessments of how individual warheads may respond to accidents; the probability and consequences of plutonium dispersion; and an extensive assessment of security problems associated with various transportation modes. It includes specific investigation of all modes of transportation (air, rail, ship, truck) and assessment of the transport containers used. The study is intended to serve as a tool for service commanders in analyzing the relative safety and security of various transportation options so as to choose the one with lowest risk. It is expected to be approved and released by late 1993 (10)

The other major recent initiative is the development of a new DOD directive for the movement of nuclear weapons. This directive has not yet been approved or implemented but would replace an older one, refer to the new transportation study, and provide improved guidance to commanders

Box 2-C—Ensuring the Safety of Nuclear Weapons

The United States has always been concerned about the possibility of an accidental or unauthorized detonation of a nuclear weapon. Warheads have, thus, been designed and built with a variety of safety features and technical obstacles to prevent these occurrences. However, to provide a credible deterrent force, the weapons must also be reliable; that is, confidence that warheads will detonate when used must be high. "Nuclear weapons surety" is the phrase used to describe these often contradictory needs.

As a result of a congressionally commissioned study, a Joint Advisory Committee on Nuclear Weapons Surety (JAC) was established in September 1992 to consolidate the work of the Departments of Energy and Defense on the issue of surety. The JAC was established under the auspices of the Federal Advisory Committee Act.

The focus of the JAC has been on the safety side of the surety equation. It is tasked with advising the Secretary of Defense, the Secretary of Energy, and the Nuclear Weapons Council on matters concerning "inadvertent detonation or plutonium dispersal."

- The JAC was officially chartered on September 4, 1992 and will operate until September 4, 1994. It meets twice a year and at other times designated by the Chairman.
- The designated Federal official for the JAC is the Assistant Secretary of Defense for Atomic Energy. The Committee is authorized to have five members, at most. It currently has five, as well as several designated alternates.
- Subcommittees and panels may be created as necessary to address specific issues. Reports from such groups are given to the JAC Chairman before being released outside the Committee.
- Meetings are announced in the *Federal Register*. If meetings are closed to the public, the Executive Secretary "will issue an annual report setting forth a summary of its activities . . . as would be informative to the public."

The estimated annual operating cost of the JAC is $168,000.

The JAC considers safety issues in all aspects of nuclear weapons. For example, missile design, propellants, and transportation scenarios are examined, in addition to the warheads themselves. The JAC has not yet commissioned any substantive analyses (11,12).

SOURCE: Joint Advisory Committee on Nuclear Weapons Surety.

in judging and selecting the lowest-risk transport mode.

Several issues related to these studies involve the need to improve transportation safety and the prospects of expediting the process. First, several experts (mostly at the national laboratories) have long advocated greater safety and security in transportation, especially for older weapons that lack some of the safety features of the newer ones (9,13,14,19). These experts have been recommending the use of accident-resistant containers and Safe Secure Transport (SST) vehicles (tractor-trailers operated by the DOE Transportation Safeguards Division). At the same time, some of the armed services have claimed that the advantages of this approach may be outweighed by other considerations, such as the need to meet schedules and maintain high security of operations. Some say that accident-resistant containers can degrade security and are not worth that price.

Much effort has gone into the analysis of weapons transportation safety, and the resulting studies and proposed directives have merit. It appears that they will be adopted and implemented soon, but some constraints remain. If the process could be moved forward, it should have significant benefits for reducing the future risk of accidents that might result in the dispersal of dangerous nuclear materials. Even though accident-risk probabilities are very low, understanding how to decrease risk further, as shown by these rigorous assessments, can help decision-makers make prudent choices and take all possible measures to increase safety.

Storage Safety

The safety of intact warheads while in storage is an issue that may merit an updated review because it is possible that certain storage depots will be maintained for the long term. Another factor involves the closing of DOD bases and other restructuring of forces that may increase logistical concerns regarding movements of weapons returned to DOE. Here again, the Defense Nuclear Agency will be challenged to maintain the best and safest balance between DOD needs and DOE capabilities. DNA keeps track of weapons in DOD custody and negotiates with DOE regarding the types and numbers to be transported to a DOE site for dismantlement in a given period. A critical factor is, thus, the ability of Pantex—both its physical capacity and its worker resources—to meet schedules without sacrificing safety or overloading any part of the process. There is a possibility that Pantex schedules will be delayed if DOE cannot adequately and quickly resolve problems related to increasing the storage capacity for plutonium pits from warheads.[6]

Because the information is classified, no detailed analysis of DOD storage capabilities, scheduled closure of facilities, or transportation capacity for specific weapons return scenarios is available for public review. Also, safety oversight systems are mainly internal to DOD. These should be especially comprehensive and rigorous in protecting public health and the environment.

DOD and DOE have recently reviewed a range of nuclear weapons safety issues and listed those that may require study (in priority order) in a joint surety plan. In this list, an issue assessment of long-term storage was given a high priority. Such an assessment study may be initiated in the near future, but as of mid-1993, no firm plans have been made.

Other Safety Issues

Within the current joint surety plan, some other safety issues are given high priority. Two of these are fire resistance enhancement and dismantlement risk. The dismantlement risk issue is relegated to DOE's attention. Fire resistance enhancement is the subject of a study that has been initiated by DOD, following a key recommendation of the Drell Report. It will characterize credible future environments for warhead exposure to fire and will explore improvements in design, operational procedures, and mitigation measures that could be employed. The study began in late 1992 and will be completed in 1995.

Another concern regarding nuclear weapons safety is whether, in the rush to demilitarize, dismantle, and eliminate these major weapons, the system will still command high-quality attention and adequate resources. Although budgets may be reduced because weapons are no longer needed, it will still be necessary to provide resources to ensure that care is exercised in protecting both individuals and the environment from these devices and materials.

CONCLUSION

In sum, a substantial nuclear weapons retirement and warhead return process is now under way by DOD based on presidential directives and other factors. Retirement decisions are made in the context of which warheads are no longer needed in the active and reserve stockpiles. When such a decision is made, each service with physical control of a weapon arranges for transportation to a continental U.S. military base, if appropriate, and then puts either the weapon or the warhead into temporary storage until it can be returned to DOE. The DOD process of weapons retirement, following the national goal of stockpile reduction, has been under way for the past few years. Issues of logistical planning, safe transportation and storage, defining overall strat-

---

[6] See chapter 3 and appendix A.

egy, and making decisions about the Nation's nuclear future are being addressed by the responsible Federal agencies but with minimal public debate or input. Continuing issues that could be addressed include developing a national consensus, defining a unified mission, and ensuring that adequate improvements in safety, security, and protection of health and the environment are carried out.

CHAPTER 2 REFERENCES

1. Allison, Graham, Ashton Carter, Steven Miller, and Philip Zelikow, "Cooperative Denuclearization—From Pledges to Deeds," Harvard University, Center for Science and International Affairs, January 1993.
2. Birely, John, Deputy Assistant to the Secretary of Defense (Atomic Energy), briefing to Office of Technology Assessment, Apr. 12, 1992.
3. Boardman, Karen, Program Engineer, Department of Energy, Albuquerque Operations Office, Weapons Programs Division, briefing to Office of Technology Assessment, May 21, 1992.
4. Claytor, Richard, Assistant Secretary for Defense Programs, Department of Energy, testimony before the U.S. Senate Committee on Armed Services, Aug. 4, 1992.
5. Defense Nuclear Agency, "The Nuclear Weapons Dismantlement Process," prepared for the Office of Technology Assessment, November 1992, and briefing of OTA staff, Feb. 19, 1993.
6. Dowis, James W., Special Assistant on Stockpile Matters to the Assistant to the Secretary of Defense (Atomic Energy), letter to the Office of Technology Assessment, June 30, 1993.
7. Graham, Douglas R., Deputy Assistant Secretary of Defense, Testimony before the U.S. Senate Committee on Armed Services on Warhead Dismantlement and the START Treaty, Aug. 4, 1992.
8. Guidice, Stephen, Assistant Manager for Office of Operations and Weapons, Department of Energy, briefing to Office of Technology Assessment on DOE nuclear weapons dismantlement, May 20, 1992.
9. Harrison, James, Sandia National Laboratories, Weapons Project Group, briefing to Office of Technology Assessment on Sandia's role in dismantlement relative to environmental safety, May 20, 1992.
10. Havlik, David J., Special Scientific Advisor to the Assistant to the Secretary of Defense (Atomic Energy), briefing to Office of Technology Assessment staff, Mar. 5, 1993.
11. Joint Advisory Committee on Nuclear Weapons Surety, Authorization, May 6, 1992; and Charter, Sept. 4, 1992.
12. Keel, Stanley, Special Assistant for Nuclear Weapons Safety to the Assistant to the Secretary of Defense (Atomic Energy), personal communication, Mar. 8, 1993.
13. Kidder, R.E., "Assessment of the Safety of U.S. Nuclear Weapons and Related Nuclear Test Requirements," Report No. UCRL-IR-109503, Lawrence Livermore National Laboratory, Dec. 10, 1991.
14. Miller, George H., Lawrence Livermore National Laboratory, testimony before the U.S. House of Representatives Committee on Armed Services, Mar. 31, 1992; see also Miller, George H., Paul S. Brown, and Carol T. Alonso, "Report to Congress on Stockpile Reliability, Role of Nuclear Testing," Lawrence Livermore National Laboratory Report No. UCRL-53822, October 1987.
15. National Research Council, *The Nuclear Weapons Complex, Management for Health, Safety, and the Environment* (Washington, DC: National Academy Press, 1989).
16. Presidential Response to Section 3142 of the FY 1992 National Defense Authorization Act on Nuclear Weapons Matters, April 1992.
17. Scheinman, Lawrence, and David A.V. Fischer, "Nuclear Weapons Dismantlement: An Opportunity to Advance the Cause of Non-Proliferation," *The Atlantic Council Bulletin,* February 1992.
18. U.S. Arms Control and Disarmament Agency, "Fact Sheet on the START II Treaty," Washington, DC, Jan. 1, 1993.
19. U.S. Congress, House Committee on Armed Services, Panel on Nuclear Weapons Safety, "Nuclear Weapons Safety Report" (Drell Report), No. 15, December 1990.
20. von Hippel, Frank, "Bombs Away," *Discover,* vol. 13, No. 4, April 1992, p. 32.

Warhead Dismantlement Programs and Plans 3

When weapons are returned by the Department of Defense (DOD) to the Department of Energy (DOE), they are transported to the DOE Pantex Plant for dismantlement. Dismantlement begins with removal of all the nonnuclear components, including the chemical high explosives (HE) that surround the nuclear materials. Dismantlement also includes the management of waste materials, which comprises such steps as separation, characterization, demilitarization, sanitization, and disposal. Some materials are temporarily stored. The storage of plutonium and highly enriched uranium is discussed in chapter 4. The final disposition of nuclear materials removed from weapons has not yet been determined.

Point

"From a global security point of view, I would argue that—the fact that the Department of Energy is doing the best it can with existing dismantlement procedures until it can finalize new ones—is the only responsible approach for them to take."

National Academy of Sciences staff reviewer of OTA report

Counterpoint

"I continue to worry that the dismantlement train has left the station probably headed in the wrong direction but running on previous policies in light of the lack of new policies."

Retired DOE manager and reviewer of OTA report

PRESENT DOE ACTIVITIES AND PLANS

Overview

Several DOE facilities are currently engaged in the warhead dismantlement process, with major activities centered at Pantex, Y-12, and Savannah River (see figure 3-1). Plutonium pits[1] are removed from warheads and temporarily stored in bunkers at Pantex. Other parts and wastes are stored, characterized, and disposed of in a variety of ways that have been developed and used by DOE in the past. Secondaries[2] are shipped to the Y-12 Plant at Oak Ridge, Tennessee for further disassembly or storage. Tritium canisters are shipped to the Savannah River Site. Figure 3-2 illustrates the steps involved in dismantlement at Pantex.

[1] A plutonium pit is the primary explosive nuclear core of the warhead package.

[2] A key (self-contained) subassembly of the nuclear warhead package.

Figure 3-1—DOE Facilities Involved in Nuclear Weapons Dismantlement

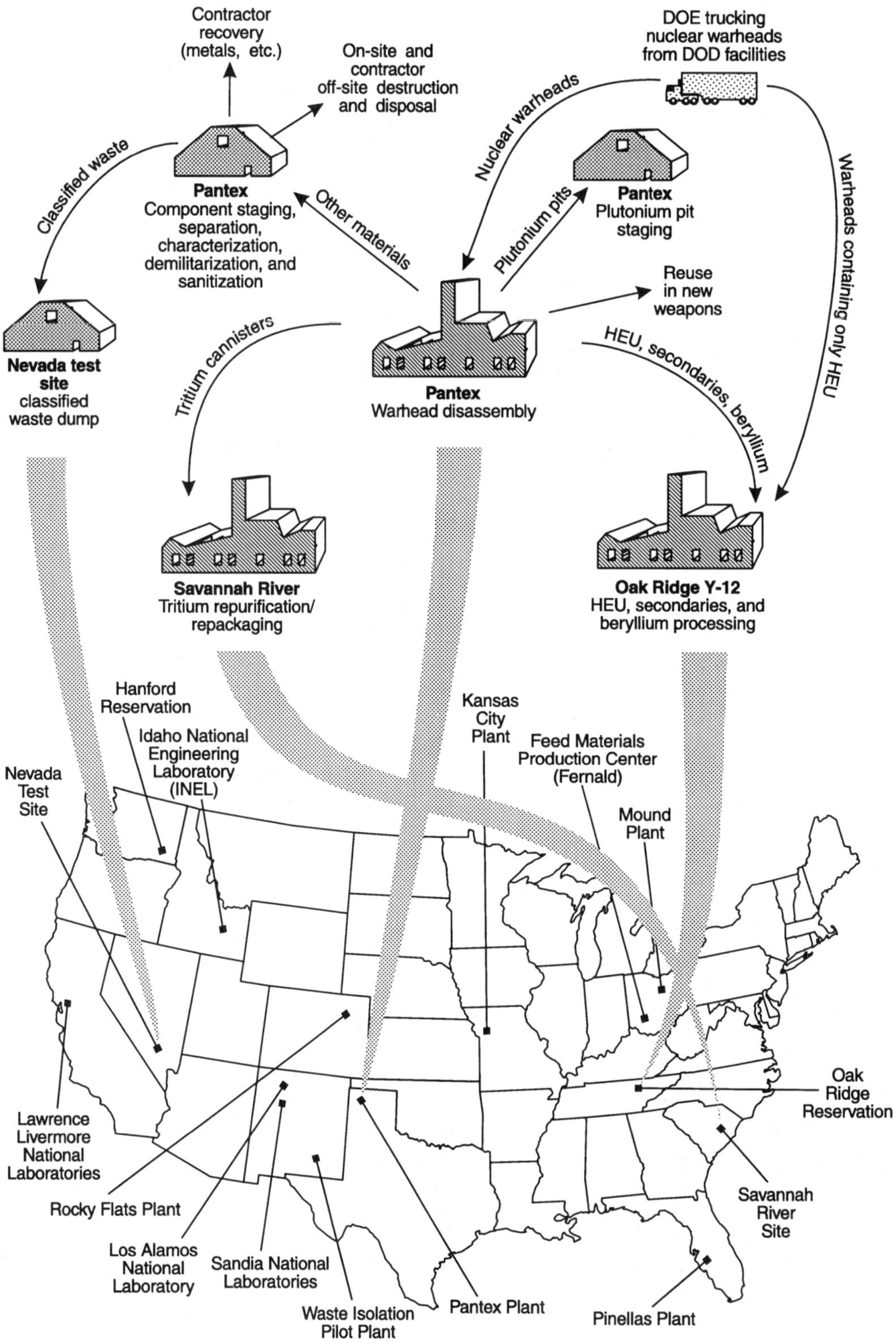

SOURCE: Office of Technology Assessment, 1993.

Figure 3-2—Warhead Dismantlement at Pantex with Flow of Parts, Materials, and Waste

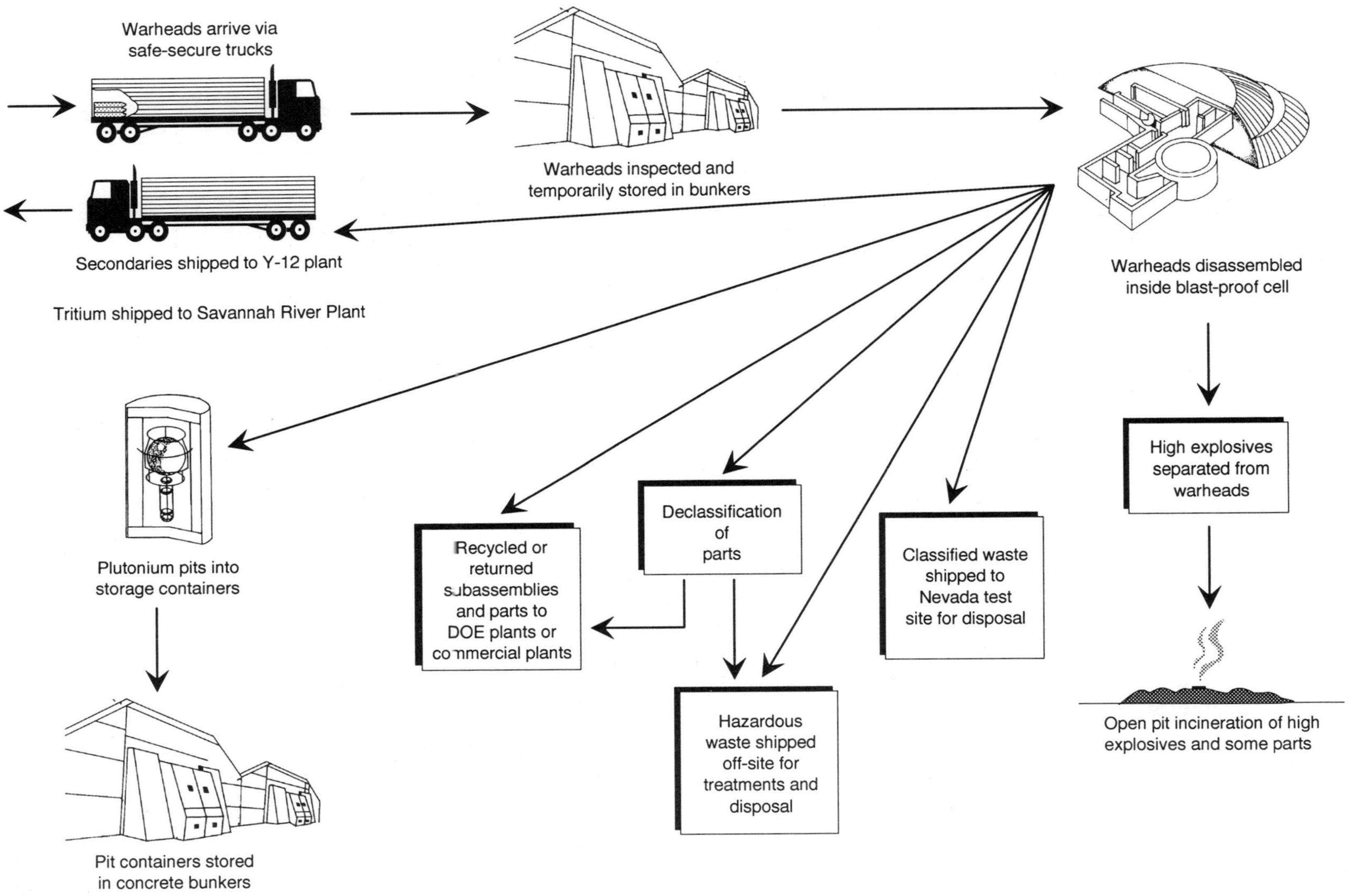

SOURCE: Office of Technology Assessment, 1993.

In the past, DOE has dismantled nuclear warheads as a part of operations to maintain the active and inactive stockpiles, and to retire obsolete weapons systems (32). However, during this earlier activity, most nuclear and nonnuclear components were returned to the DOE plants in which they were made. Now, some facilities have ceased operations and can no longer accept materials. For example, plutonium pits can no longer be returned to Rocky Flats, so these pits must be stored at Pantex. The focus of DOE's activities has changed from warhead production to warhead dismantlement. During FY 1991, 1,546 warheads were dismantled at Pantex. In FY 1992, even though the total number dismantled by DOE was much larger, the number at Pantex decreased to 1,274 because one type was of such design that it could be sent directly to the Y-12 Plant and dismantled there. As of summer 1993, DOE estimates that about 1,400 warheads will be dismantled at the Pantex Plant during FY 1993 (83). The pace of dismantlement and the difficulty of dismantlement work are affected by management, political, and technical challenges.

Management challenges involve integrating environmental, safety, and health improvements into dismantlement operations. DOE has asserted its commitment to protection of the environment, safety, and health (89). This commitment represents a distinct change from the traditional DOE culture, and efforts to establish a "new culture" have been spurred in large part by recognition of the widespread contamination resulting from past DOE production practices and revelations about safety hazards at DOE facilities. Environmental, safety, and health practices at Pantex and other plants involved in dismantlement have been strongly criticized by both internal and external reviews (1,17,23,59,67,68,82). Improvements have been noted in followup reviews, yet much remains to be accomplished (16,84). Another management challenge is to maintain an aggressive schedule of warhead dismantlement while solving logistics problems in the safest practical way. DOD now has a backlog of retired weapons and

U.S. DEPARTMENT OF ENERGY

A Pantex Plant worker removes a nuclear warhead from a shipping container.

has assigned priority ranking to different types for return to DOE. DOE management must serve DOD needs and handle particular weapons systems without sacrificing safety, efficiency, and effectiveness.

Political challenges include the increased interest of the States and the public in activities at the Weapons Complex. Recent requirements for DOE to comply with State environmental regulations provide greater opportunities for States and public interest groups to oversee DOE operations, and give these groups additional leverage in affecting dismantlement activities. For example, the State of Texas is responsible for hazardous waste at Pantex under the Resource Conservation and Recovery Act (RCRA) and for issuing air quality permits governing emissions associated with the burning of HE at Pantex. However, fragmented regulatory responsibilities among different State and Federal agencies, as well as lack of jurisdiction over some materials, may limit oversight.

Technical challenges associated with dismantlement stem from the complexity of nuclear warheads. There are many components in a warhead—including nuclear materials, other toxic

Figure 3-3—Materials Generated from Dismantling a Typical Nuclear Warhead

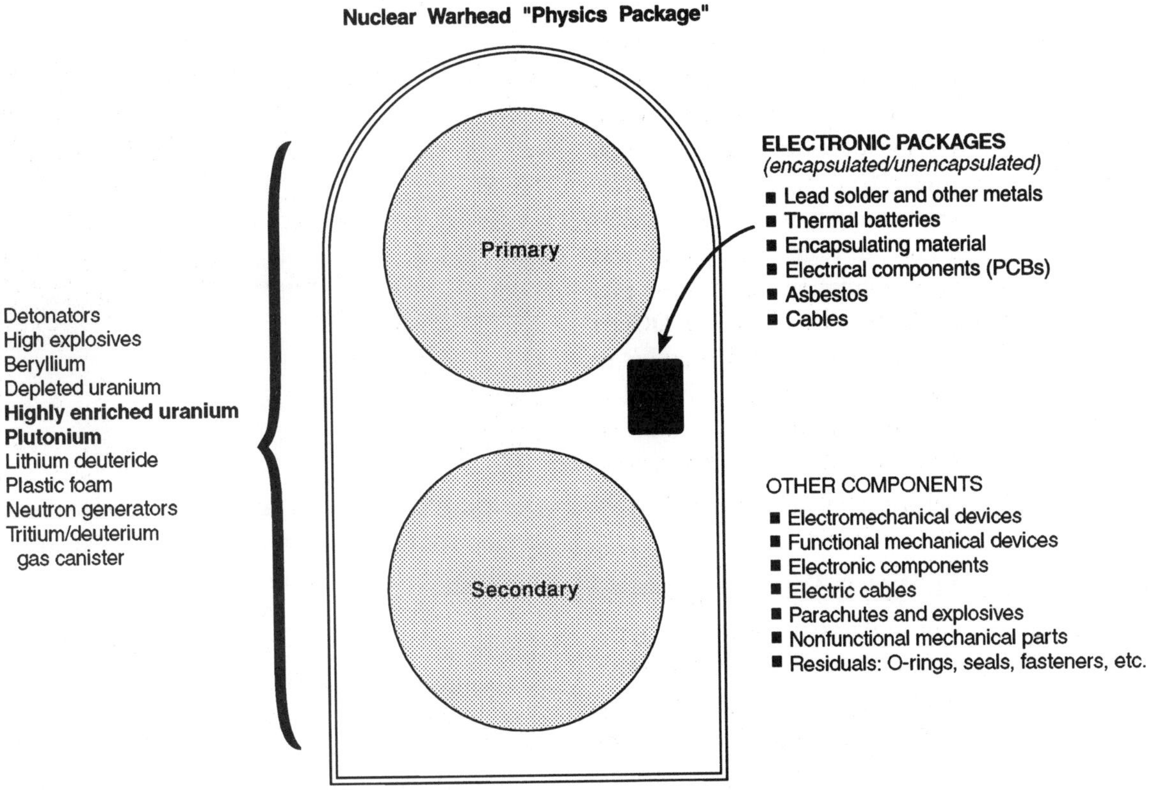

SOURCES: Adapted from briefing by Johnny Grant, Project Leader for Nuclear Weapons Retirement and Disposal for the U.S. Department of Energy (Mar. 17, 1992); and from the National Research Council, *The Nuclear Weapons Complex, Management for Health, Safety, and the Environment* (Washington, DC: National Academy Press, 1989).

and hazardous materials (chemicals and metals), and classified materials (switches, electronic components)—all of which require careful handling and attention to environmental, safety, and health issues (see figure 3-3 for a list of materials contained in nuclear warheads). Although weapons have engineered safety features to prevent inadvertent detonation, some of the explosives in the older warheads are sensitive to shock and require controlled conditions. Newer weapons are more resistant to unintentional detonation of the explosives. Most, if not all, weapons slated for dismantlement are older, more sensitive warheads, as are some weapons that are still deployed.

Finally, each weapons type is unique, requiring different dismantlement tools and procedures, and possibly different methods of component disposal. Each change in weapons type requires extensive planning, operations reviews, safety reviews, worker training, and special tests. The time required for these can vary significantly depending on the type and design of weapon. DOE has estimated that the time required for only the main disassembly tasks for several warheads could vary by a factor of four (83). Information on whether certain components, such as capacitors, contain hazardous materials is not always available to DOE. Thus some materials will have to be tested to identify the chemicals present, prior to dismantlement and disposal.

Although most recent problems with dismantlement appear to be minor, some have been of particular concern and have affected DOE's schedules. In April 1993, DOE reported that the outer layer of a plutonium pit had cracked open during disassembly of a W48-type warhead. The actual incident occurred in November 1992. The pit had to be removed from the normal process for examination and testing. DOE stated that the incident resulted in minor contamination of the work area. Since this related to one specific weapons type, further work on that system was discontinued until investigations could be completed and new procedures developed. Because of the time required to shut down one line and start up another, this type of problem can have a significant effect on schedules and rates of dismantlement. Any future problem can be expected to have similar effects (24,78).

Steps and Facilities in Dismantlement

Nuclear warheads are disassembled into components at the Pantex facility, in Amarillo, Texas. Parts from the weapons are, in general, returned to the facility where they were originally produced. An exception is that plutonium pits remain at Pantex in temporary storage, rather than being returned to Rocky Flats, which has been closed. Chemical explosives removed from a warhead's pit are burned at Pantex. Individual plutonium pits are put in steel containers resembling oil drums, which are then placed in earth-covered concrete bunkers.

Warheads arrive at Pantex's Shipping and Receiving Building in specialized vehicles, and undergo several inspections and safety checks. They are then unloaded at 1 of 60 storage facilities (bunkers) contained in the secured area about 1 mile from the assembly/disassembly area (91). As of March 1992, approximately 42 bunkers were used for weapons or weapons component staging (51).

The steps required to assemble nuclear warheads are reversed for dismantlement (79). Thus, after conducting several inspections and safety checks on a warhead, disassembly personnel remove the warhead's cover followed by all

U.S. DEPARTMENT OF ENERGY

Aerial view of the Pantex Plant, near Amarillo, Texas.

electrical components and other hardware. Inspection of warheads includes the use of radiography to verify the configuration and condition of warhead components. Additional tests may be conducted to determine, among other characteristics, actual mass properties, dynamic balance, and center of gravity (91).

Removal of the protective case containing high explosives and nuclear materials (e.g., physics package) is generally followed by actual separation of the nuclear and HE components. To reduce the potential for large radionuclide releases in case of the accidental explosion of conventional chemical high explosives (used in most of the U.S. nuclear stockpile), these activities are conducted in assembly/disassembly cells or "Gravel Gerties" (79). Most of the 13 assembly cells found at Pantex are used, or could be used, for disassembly of the physics package (91). Weapons containing insensitive high explosives may be assembled or disassembled in "bays," which do not have the same level of explosive containment as cells.

The disassembly of weapons results in parts containing HE, special nuclear materials, components containing hazardous and nonhazardous chemical constituents, and certain other materials that, because of the classified nature of their design, must be declassified or "sanitized" prior to treatment and disposal. Weapons disassembly also generates solvents, classified metal components, and other regulated hazardous materials (75). A variety of hazardous wastes may be generated by this process, and their proper management (treatment, storage, and disposal) represents a technical challenge.

Pantex bunkers are used for temporary storage of special nuclear materials. Most radioactively contaminated wastes and classified or nonclassified weapons components gathered during disassembly are shipped off-site for treatment and disposal (91). High explosives are burned at Pantex (79,91). The majority of recoverable material generated during the weapons dismantlement process is shipped to commercial vendors for recycling (91). Commercial waste handling facilities are used to handle the off-site disposal of nonnuclear waste from warhead dismantlement (5,61).

The Y-12 Plant at the Oak Ridge Reservation in Tennessee is a major weapons component manufacturing facility. Built in 1943, the plant initially separated fissile uranium-235 from natural uranium. As the Nation's weapons programs changed over the years, so did the capabilities of Y-12. Lithium separation became a mission of the plant in the 1950s. Presently, Y-12 has facilities to fabricate weapons components from uranium, beryllium, and lithium, and the plant has played a major role in producing nearly every nuclear weapon in the Nation's arsenal.

Secondaries from warheads disassembled at Pantex are shipped to Y-12, and many of them are now stored intact. The Y-12 Plant receives highly enriched uranium from disassembled weapons. Other components from warheads are also shipped to the site. Uranium is stored at Y-12, and other parts are either stored or treated as waste and disposed. Some fabrication continues, however, and Y-12 currently supports the DOE laboratories, nuclear reactor projects, and the Navy's Nuclear Submarine program.

DOE's Savannah River Site in South Carolina recycles tritium, a radioactive isotope of hydrogen used to boost the explosive yield of nuclear weapons. In the past, the Savannah River Site also produced tritium and plutonium. Tritium from dismantled weapons is stored at Savannah River and purified for reuse. Because of its short half-life, tritium must be resupplied to weapons.

Management and Oversight Structure for Dismantlement

Dismantlement is currently the responsibility of the DOE Office of Defense Programs (DP). Actual warhead dismantlement is conducted by the management and operations (M&O) contractor, Mason & Hanger-Silas Mason Co., Inc. (M&H), for the Pantex Plant. The DOE Amarillo Area Office is located at Pantex and reports to the Albuquerque Operations Office. The Albuquerque Operations Office reports to the DOE Assistant Secretary for Defense Programs. Technical support is provided by several of the national laboratories that originally designed the warheads now scheduled for dismantlement. DP has responsibility for developing and implementing environmental, safety, and health policies for its operations. DP receives policy guidance and oversight from other DOE headquarters offices such as the Office of Environment, Safety, and Health (EH) (85).[3]

To address new challenges in dismantlement operations, on December 31, 1991, DOE established a special task force, the Executive Management Team for Dismantlement (EMTD), with broad oversight of nuclear warhead dismantlement operations (62). The Albuquerque Operations Office was designated to establish and chair EMTD (5). EMTD activities include: 1) establishment of materials identification and disposal teams; 2) characterization of materials for disposal; 3) development and procurement of specific tools; 4) identification and resolution of

[3] EH now includes the formerly independent Office of Nuclear Safety.

environmental, safety, and health concerns; 5) definition of treatment and disposal methods for materials from weapons dismantlement; 6) updating retirement disposal instructions; 7) detailing specific operating procedures; 8) training operators; 9) evaluating nuclear safety; and 10) conducting a final program review (5).

Despite the preliminary stage of some EMTD activities, several Weapons Complex facilities are undertaking warhead dismantlement tasks using existing operating procedures, in part because DOE considers weapons dismantlement to be a logical extension of past operations. According to this view, dismantlement has merely changed the emphasis from assembly to disassembly, but the same techniques, personnel, facilities, and skills are involved.

Although EMTD is addressing new challenges in Pantex dismantlement operations, DOE still relies on many conventional methods. It continues to dispose of components or materials from dismantled warheads as usual, rather than waiting until new methods are developed (44). EMTD recommendations are integrated into ongoing operations only if the site operations manager determines a change is needed.

Internal DOE oversight functions were changed during the tenure of former Secretary of Energy James D. Watkins, and new policies and guidance are being developed (85). It has been difficult in the past to ensure that environmental, safety, and health guidelines were being followed, and the internal oversight office (EH) currently has insufficient mechanisms to require specific compliance or to enforce its requirements. In addition, EH lacks the personnel to review progress at field offices. Hiring personnel with expertise in occupational safety and health has been difficult not only for the oversight office but for DOE line organizations as well.

The Cost-Plus Award Fee (CPAF) process was established to help increase emphasis by M&O contractors on environmental, safety, and health factors. This process provides a mechanism for evaluating progress on meeting some defined objective, but there is little evidence that it actually increases management's attention to health and safety issues (65).

External advisory oversight, with particular emphasis on nuclear safety issues (e.g., criticality safety, training of radiation workers) is provided by the Defense Nuclear Facilities Safety Board (DNFSB). Regulatory oversight on environmental matters at Pantex is provided by the Environmental Protection Agency (EPA) and the State of Texas. Similar management, contractor, and oversight arrangements prevail at other dismantlement sites (e.g., Y-12 in Tennessee, Savannah River in South Carolina).

The DNFSB is a relatively new external advisory oversight mechanism. It was created by Congress in 1989 to provide advice and recommendations to the Secretary of Energy on public health and safety at DOE defense nuclear facilities (12). In 1991, Congress amended the enabling act and broadened its jurisdiction to include the assembly, disassembly, and testing of weapons, thus expanding DNFSB oversight to Pantex operations. DNFSB reviews facilities, operations, practices, and occurrences at DOE facilities. It examines the safety practices of both DOE and the M&O contractors. DNFSB also reviews and evaluates the content and implementation of health and safety standards, including DOE orders, rules, and other safety requirements. Table 3-1 lists the orders subject to DNFSB oversight.

DNFSB's primary tool for gaining the attention of DOE is to issue recommendations to the Secretary of Energy that require a response or a report. Some DNFSB recommendations are specific to DOE facilities (e.g., recommendations 91-2, Closure of Safety Issues Prior to Restart of the K-Reactor at the Savannah River Site; and 90-7, Safety at Single-Shell Hanford Waste Tanks). Others apply broadly to all DOE defense nuclear facilities (e.g., 91-1, Strengthening the Nuclear Safety Standards Program for DOE's Defense Nuclear Facilities; 91-6, Radiation Protection for Workers and the General Public at DOE Defense Nuclear Facilities). Recommendation 91-6 (11)

Table 3-1—DOE Orders Subject to DNFSB Oversight

Part I. Weapons-Sensitive DOE Orders

Order number	Subject
5530.1A	Response to Accidents and Significant Incidents Involving Nuclear Weapons
5530.2	Nuclear Emergency Search Team
5530.3	Radiological Assistance
5530.4	Aerial Measuring
5600.1	Management of DOE
5610.1	Weapons Complex Packaging and Transportation of Nuclear Explosives, Nuclear Components, and Special Assemblies
5610.10	Nuclear Explosives and Weapons Safety
5610.11	Nuclear Explosive Safety
5610.13	Joint DOE/DOD Nuclear Weapons System Safety, Security, and Control

Part II. Safety-Related DOE Orders

Order number	Subject
1300.2A	Department of Energy Technical Standards Program
1360.2A	Unclassified Computer Security Program
1540.2	Hazardous Materials Packaging for Transport–Administrative Procedures
1540.3	Base Technology for Radioactive Material Transportation Packaging Systems
1540.4	Physical Protection of Unclassified, Irradiated Reactor Fuel in Transit
4330.4A	Maintenance Management Program
4700.1	Project Management System
5000.3A	Occurrence Reporting and Processing of Operations Information
5400.1	General Environmental Protection Program
5400.2A	Environmental Compliance Issue Coordination
5400.3	Hazardous and Radioactive Mixed Waste Program
5400.4	Comprehensive Environmental Response, Compensation, and Liability Act Requirements
5400.5	Radiation Protection of the Public and the Environment
5440.1D	National Environmental Policy Act Compliance Program
5480.1B	Environmental, Safety and Health Program
5480.3	Safety Requirements for the Packaging and Transportation of Hazardous Materials, Hazardous Substances, and Hazardous Wastes
5480.4	Environmental Protection, Safety and Health Protection Standards
5480.5	Safety of Nuclear Facilities
5480.6	Safety of DOE-Owned Reactors
5480.7	Fire Protection
5480.8	Contractor Occupational Medical Program
5480.9	Construction Safety and Health Program
5480.10	Contractor Industrial Hygiene Program
5480.11	Radiation Protection for Occupational Workers
5480.15	DOE Laboratory Accreditation Program for Personnel Dosimetry
5480.17	Site Safety Representatives
5480.18A	Accreditation of Performance-Based Training for Category A Reactors and Nuclear Facilities
5480.19	Conduct of Operations Requirements for DOE Facilities
5480.20	Personnel Selection, Qualification, Training, and Staffing Requirements at DOE and Nonreactor Nuclear Facilities
5480.21	Unreviewed Safety Questions
5480.22	Technical Safety Requirements
5480.23	Nuclear Safety Analysis Reports
5481.1B	Safety Analysis and Review
5482.1B	Environment, Safety, and Health Appraisal Program
5483.1A	Occupational Safety and Health Program for DOE Contractor Employees at Government-Owned Contractor-Operated Facilities
5484.1	Environmental Protection, Safety, and Health
5500.1B	Emergency Management System
5500.2B	Emergency Categories, Classes, and Notification and Reporting Requirements
5500.3A	Planning and Preparedness for Operational Emergencies
5500.4	Public Affairs Policy and Planning Requirements for Emergencies
5500.7B	Emergency Operating Records Protection Program
5500.10	Emergency Readiness Assurance Program
5700.6C	Quality Assurance
5820.2A	Radioactive Waste Management
6430.1A	General Design Criteria

SOURCE: Defense Nuclear Facilities Safety Board, 1992.

led to the development of DOE's new Radiological Control Manual (RADCON) (80), discussed later in this chapter.

Another mechanism used by DNFSB to effectuate change at DOE facilities involves the discussions and technical exchanges that are an integral part of site visits. The interaction between DNFSB and DOE/contractor staffs often leads to improvements that are not reflected in formal recommendations. This mechanism and the lack of public access to the Board's work have been criticized. Some critics claim that the Board is not truly independent—that DNFSB lacks adequate authority because it merely advises the Secretary of Energy and has never really solicited broad public input (9,55).

On the other hand, DNFSB identifies a variety of examples of public participation, including correspondence with STAND (Serious Texans Against Nuclear Dumping) and the State of Texas in response to concerns about staging configurations for special nuclear materials and other safety issues at Pantex, as well as public access to DNFSB recommendations, annual reports, notices of public meetings/hearings, and material available after public meetings in accordance with Sunshine Act and Freedom of Information Act rules. These rules were developed in response to litigation.[4] Results of this litigation are that DNFSB has complied with these regulations but that meetings involving formal recommendations to the Secretary of Energy or the President may be closed (21). The Supreme Court declined to review this case in a ruling on May 17, 1993 (90).

Many DNFSB recommendations and site visits focus on increasing the formality of written procedures and directions in DOE operations and in its training of workers. This emphasis may be a reflection of the background of many DNFSB staff in commercial and naval nuclear reactors. The DNFSB advisory role has been taken very seriously by DOE. All recommendations, to date, have been accepted. Some commentators within DOE have expressed concern that this effort to satisfy DNFSB may divert attention from more comprehensive needs. They claim that the resources needed to improve overall training activities are sometimes directed into the narrower areas identified by DNFSB. These individuals contend that DNFSB plays a very powerful, almost regulatory, role, yet its recommendations are not subject to the outside review and scrutiny faced by regulatory bodies. For example, the Nuclear Regulatory Commission has formal mechanisms for rulemaking and public comment, including publishing *draft* recommendations in the *Federal Register*. DNFSB recommendations are issued in *final* form for public comment.

Between May 1992 and June 1993, DNFSB conducted numerous site visits at weapons dismantlement facilities. Most of this effort focused on Pantex, with additional visits to Y-12 and Los Alamos National Laboratory, as well as some review of the DOE Amarillo Area Office and the Albuquerque Operations Office. The visits resulted in correspondence with DOE concerning various safety issues, including training, procedures, conduct of operations, compliance with orders, safety analysis, criticality safety, dosimetry, Operational Readiness Review process, radiation control practices, and safety of nuclear explosives. DNFSB has required reports on criticality safety at Pantex (17) and radiation control practices, and plans to remedy DOE order compliance deficiencies at Y-12 (23). DNFSB has also made specific recommendations applicable to dismantlement facilities (Recommendation 93-1, Standards Utilization in Defense Nuclear Facilities) (19).

---

[4] See *Energy Research Foundation* v. *Defense Nuclear Facilities Safety Board,* 734 F. Supp. 27 (D.D.C. 1990); *Energy Research Foundation* v. *Defense Nuclear Facilities Safety Board,* 917 F.2d 581, 585 (D.C. Cir. 1990); *Natural Resources Defense Council/Energy Research Foundation* v. *Defense Nuclear Facilities Safety Board,* No. 91-1199 (D.C. Cir., July 24, 1992); *Natural Resources Defense Council/Energy Research Foundation* v. *Defense Nuclear Facilities Safety Board,* No. 91-1199 (D.C. Cir., Oct. 9, 1992).

Dismantlement Procedures

DOE management and the national laboratories jointly produce the standard operating procedures for nuclear weapons dismantlement at Pantex (32). However, the laboratories have final approval authority. The national labs have designed all U.S. nuclear weapons. Consequently, Sandia, Lawrence Livermore, and Los Alamos constitute a unique source of information about the nuclear weapons slated for dismantlement (34).

Some procedures for reviewing and approving dismantlement activities for particular weapons programs are in place. The three principal steps involved are: 1) Operational Readiness Review, 2) Operational Readiness Evaluation, and 3) Qualification Evaluation for Dismantlement Release. These procedures are internal to DOE and do not involve public or outside scrutiny. However, DNFSB has taken an active role in reviewing this process at Pantex and other DOE facilities.

The Operational Readiness Reviews (ORRs) are conducted by a team of contractor engineers at the dismantlement facility (Pantex or Y-12). These internal reviews are intended to ensure that the procedures and equipment necessary to begin dismantlement operations are in place. After the ORR, DOE conducts an Operational Readiness Evaluation (ORE), which is a critique of the ORR and confirms whether the activity is ready to go. The Qualification Evaluation for Dismantlement Release (QED) is an additional review by national laboratory design engineers to verify the DOE critique. This step was added as a result of concerns expressed by DNFSB about the ORR-ORE process for a particular weapons system at Pantex. In that example, the ORE found the ORR to be insufficient. DNFSB conducted an independent review of the ORE, and noted deficiencies in the process used and in safety factors (18). In particular, DNFSB expressed concern about continuing deficiencies in the Safety Analysis process at Pantex, an issue that has been raised repeatedly by both internal and external reviewers since 1989. DNFSB had also criticized the ORR-ORE process at Pantex in July 1992 (14).

U.S. DEPARTMENT OF ENERGY

Pantex Plant worker begins warhead disassembly.

In practice, many of the informal mechanisms used in the past to guide M&H dismantlement operations are still in effect. M&H engineers meet with their counterparts at the national laboratories to discuss technical challenges. Laboratory personnel then work on solutions to these problems. If promising technologies are developed, it is often up to M&H to determine whether or not to pursue a new approach, unless it is ordered to make a change by the DOE Albuquerque Operations Office. Some national laboratory personnel who have worked closely with M&H are part of EMTD; others are not.

Recent work on the use of robotics in dismantlement operations has been supported by EMTD. Some efforts have explored the application of robotics to reduce human radiation exposure from dismantlement activities. These methods are being evaluated by Sandia National Laboratory (see box 3-A). They are examples of some of the dismantlement process design work sponsored by DOE. There is no clear connection, however, between such design work and any overall assessment of dismantlement technology needs within the Pantex management organization.

Box 3-A—Using Robotics to Improve the Dismantlement Process

Although the complete dismantlement of nuclear warheads by robots is unlikely, they could be used in certain steps in the dismantlement process. Robots have been developed and used to improve commercial nuclear reactor safety through reducing human radiation exposure and operations in inaccessible areas (93). However, it is unlikely that humans will be completely replaced in warhead dismantlement operations at Pantex because safety considerations dictate that the initial handling and disassembly operations be done manually (39). Nuclear warheads were designed to be assembled and disassembled by humans, and therefore have idiosyncratic variations—such as the positioning of wire bundles—that make robotics programming very complicated and continue to require humans for the safety of some operations (93).

To evaluate opportunities for robotics to improve warhead dismantlement operations, Sandia National Laboratory conducted studies to identify the most risky parts of dismantlement. The Sandia robotics group did risk and exposure assessments and then developed robotics systems to replace human personnel in the more dangerous operations (35). One study looked at issues including the use of remote sensing, verification, and specialized robotics for nuclear warhead components staged in bunkers at Pantex (35).

DOE and the national laboratories have identified weapons disassembly; the handling, storage, and transportation of nuclear materials; and nuclear materials monitoring, accountability, and inventory as areas in which robotics could play a part, largely by using existing commercial technology (39).

Robotics could reduce human radiation exposure during dismantlement: one example is the packing and unpacking of plutonium pits in steel storage containers. This occurs several times during dismantlement of a single warhead and represents a substantial proportion of the exposure risk (39).

Other examples in which robotics may be useful are the removal of detonators; the removal of chemical high explosives; pit processing; "swiping" (surface sampling) the pit to detect radioactive contamination; cleaning; weighing the pit; and checking for leaks (39). One robotics system under development accepts the warhead explosive assembly, cuts the explosive charge by using high-pressure water jets, applies hydraulic pressure to separate the pieces, and packages the remains for storage—all with remote human supervision (93).

A specific example of Sandia's robotics projects is "Project Stage Right," with involvement of contractors Mason & Hanger and Battelle. The project is designed to avoid worker radiation exposure from regular inventory inspections, which would increase under DOE's proposed double plutonium pit stacking plan (7). DOE is interested in avoiding human inventory inspection by allowing remote inspection via robotics (31). DOE's Pantex double pit stacking plan envisions staging and regular monitoring of 15,000 plutonium-pit-containing steel drums in 38 bunkers (7). The project will reduce worker radiation exposure from plutonium pits during loading and unloading, handling, and inventory operations through the use of manned, radiation-shielded forklifts to handle pallets of steel drum pit containers; an unmanned remotely controlled autonomous forklift that can handle and inventory pallets; and an automatic system for inventorying plutonium pits on a regular basis, with or without physical entry into the bunker (7).

SOURCE: Office of Technology Assessment, 1993.

Costs of Dismantlement

Because there is no requirement or procedure for separating dismantlement costs from other production and surveillance costs within DOE Defense Programs, very little integrated information is currently available that reflects the total cost of this new mission. The DOE Albuquerque Operations Office has, however, responded to Office of Technology Assessment (OTA) requests for available dismantlement cost data. In FY 1993, direct costs are estimated to be about $25 million, and in FY 1994 in excess of $30 million. At least two-thirds of these costs are attributed to Pantex activities and about one-eighth to Y-12. Estimates represent direct costs only and do not include many other items that would need to be considered, such as security, maintenance, and oversight activities. For example, in FY 1993, operation of the weapons and materials transportation system by DOE is expected to cost about $79 million, most of which is used for dismantlement activities. This system is used to transport all warheads from military bases to the Pantex Plant and secondaries from Pantex to Y-12 (42,43,75,76,92).

Such cost figures probably fall far short of the totals for the comprehensive mission of dismantlement. The total operating budgets for the two sites most engaged in dismantlement—Pantex and Y-12—are $240 and $460 million, respectively, for FY 1993. Managers at each of these sites have said that at least two-thirds of their current efforts are devoted to the dismantlement mission or related work. Thus, current expenditures for dismantlement activities at these two sites alone would be almost half a billion dollars. If work at other sites, research and support activities at the weapons design laboratories, transportation and security, and oversight and management efforts are included, the FY 1993 DOE budget allocated to warhead dismantlement and materials disposition is in the range of $500 million to $1 billion (42,43,75,76,92).

Future costs of the dismantlement and management of weapons materials are unknown at this time. DOE is now preparing a plan for reconfiguration of the Weapons Complex that will likely incorporate these activities. If new facilities are to be built, their capital costs must also be considered. Such costs would need to include the provision of improved health and safety conditions for workers, as well as improved waste management practices that might accompany state-of-the-art facilities. Perhaps DOE will include estimates of dismantlement and materials management costs as part of its reconfiguration plan. If so, the increase in future costs could be evaluated in light of the need to replace many old facilities. However, there is no firm indication that construction of new facilities will be part of DOE's plan, and some contend that new facilities are not needed (30). In addition, costs to clean up the current contamination and dispose of wastes that accumulated from past practices should be factored into overall dismantlement costs.

ENVIRONMENTAL, HEALTH, AND SAFETY ISSUES IN DISMANTLEMENT

Numerous environmental, safety, and health deficiencies have been identified at Pantex and other dismantlement facilities during the past few years, and efforts are under way to address these problems. As discussed below, some outside observers are critical of past practices and skeptical of DOE's ability to significantly improve the environmental, safety, and health aspects of its operations. Certain improvements have been made, however, and changes in DOE's management approach could bring results.

Health and Safety Issues

Specific steps, processes, materials descriptions, and other aspects of dismantlement work are classified. Only general outlines of operations and worker responsibilities can be described without citing classified information. In this unclassified study, OTA did not review specific dismantlement procedures to evaluate worker hazards. OTA analysis of health and safety issues is based on unclassified information and data associated with warhead dismantlement, general principles of occupational and radiological health and safety, and unclassified evaluations of health and safety programs at dismantlement facilities by internal and external oversight groups. Box 3-B describes some basic, unclassified facts about dismantlement related to worker health and safety.

Standards and operating procedures governing the health and safety of workers undertaking dismantlement operations are developed and enforced by DOE. DOE is exempted from regulatory oversight by the Occupational Safety and Health Administration (OSHA). However, DOE is planning a transition to external regulation by OSHA over the next 3 to 5 years (72). New guidance from DOE is being developed in nonnu-

Box 3-B—Worker Health and Safety Issues in Current Dismantlement Processes

- Dismantlement occurs in highly secure cells and bays.
- Dismantlement workers handle a variety of materials that can potentially cause health and safety problems, including radionuclides, explosives, and toxic metals or organics (see figure 3-3, list of weapons components).
- Many warheads slated for disassembly lack modern safety devices, which might increase the risk of accidental detonation. The disassembly bays (Gravel Gerties) are designed to contain any plutonium released if an accidental detonation were to occur, but such an accident could have severe consequences for workers inside the bays.
- Each weapons system requires different procedures and unique tooling.
- In some procedures, workers must rely on personal protective equipment such as respirators, rather than preferred methods of engineering controls, for protection.
- Different processes are used to remove explosives from pits, some involving solvents, hydrojets, and/or thermal treatment.
- Workers are trained and certified for dismantlement work.
- Review of operating procedures by safety and health experts is a relatively new measure.
- Health and safety are governed by DOE and implemented by contractors.
- Employees have been unexpectedly exposed to radiation and hazardous chemicals.
- Unexpected radiological contamination has been detected during routine monitoring activities.
- Employees have failed to follow correct procedures.
- Work has stopped because of employee uncertainty about procedures.

SOURCE: Office of Technology Assessment, 1993.

clear health and safety areas, as well as radiation protection. The new rules are an outgrowth of numerous critical reviews of environmental, safety, and health practices at DOE facilities by both internal and external oversight groups (e.g., Tiger Teams, Advisory Committee on Nuclear Facility Safety, Defense Nuclear Facilities Safety Board).

Key issues for worker health and safety are whether efforts by DOE to exercise health and safety oversight responsibility are adequate and whether sufficient resources are being devoted to worker protection. Various reviews of DOE and contractor performance on health and safety issues are discussed below along with new initiatives to improve radiation control programs at Pantex and Y-12.

PANTEX PLANT

During dismantlement, workers must handle a variety of materials, some of which are toxic, hazardous, and/or radioactive. Box 3-C lists examples of unclassified exposure incidents involving dismantlement activities at Pantex over the past few years.

Numerous health and safety problems at Pantex were identified by the Tiger Team sent there from DOE headquarters in October 1989, including:

- inadequate radiation protection;
- inadequate hazard identification and communication;
- insufficient resources to accomplish environmental, safety, and health goals;
- inadequate policy guidance toward these goals; and
- insufficient management attention to environment, safety, and health (68).

This investigation indicated that compliance was difficult to evaluate, given the confusing array of DOE requirements and the lack of routine operations at the site during the field investigation. Responses to specific problems identified in the Tiger Team report are documented in the 1990

Box 3-C—Examples of Unclassified Worker Exposure Incidents at Pantex During Dismantlement

Beryllium Incident

According to a Tiger Team assessment, the Occupational Safety and Health Administration (OSHA) has documented an example of appropriate attention to health and safety matters involving employee exposure to hazardous material. The Pantex Industrial Hygiene Department identified employee exposures to beryllium during cleanup operations at Firing Site 23 that exceeded the OSHA Permissible Exposure Limit (PEL). Its response was to ensure the use of appropriate personal protective equipment (PPE) followed by construction of a permanent decontamination chamber (under construction during the OSHA visit). Remonitoring was planned to determine the effectiveness of a decontamination chamber in reducing employee exposure levels. Pantex management will consider further engineering controls, if necessary to meet the PEL (68).

Depleted Uranium-238 and Other Metal Residues

During the Tiger Team visit, OSHA investigated an exposure incident that occurred on February 28, 1989, about which the Metal Trades Council was concerned (68). The incident occurred during a disassembly operation. An employee was determined to have alpha contamination on his coveralls. Visible contamination in the bay was identified as depleted uranium-238, lead, aluminum, chromium, manganese, nickel, zinc, and cadmium. The Council concerns included: 1) the lack of instructions to employees on the use of personal protective equipment such as respirators; 2) inadequate biological monitoring in that only some of the employees who were potentially exposed were provided blood lead monitoring, and then only several months after the incident; and 3) the delay in providing whole body radiation counts to the affected employees.

OSHA's investigation revealed inadequate instructions and oversight in the use of PPE for the disassembly activity; inadequate monitoring of metals and radiation; inappropriate work practices involving PPE, which may have resulted in additional contamination of areas both inside and outside the disassembly bays; and inadequate recordkeeping of monitoring results. OSHA was unable to follow up its investigation by conducting air monitoring of a similar disassembly operation, which would be standard practice in investigating this type of complaint, because this sort of disassembly operation was not scheduled during the OSHA visit. OSHA did not comment about the Council's concerns regarding medical surveillance response to this incident.

Radiation-Contaminated Scrap Parts

Unexpected radiation contamination of scrap parts of weapons was detected during routine monitoring of a shearing step to declassify the parts on April 23, 1992 (77). Operations were stopped, access to the area was controlled, personnel were checked for contamination, and none was detected. The area was monitored for radiation, and contamination was found on the shearing equipment. The material was followed upstream and traced to the Burning Grounds, where contamination was located and removed. Additional surveys were conducted with no detection of further contamination. The reason for the contamination was not identified. However, steps are being taken to minimize the impact of similar incidents in the future through early detection and control of contamination at the source by additional monitoring of the Burning Grounds.

SOURCE: Office of Technology Assessment, 1993.

Action Plan prepared by Mason & Hanger and the corrective actions tracking system that includes a formal certificate of completion. The effectiveness of some corrective actions documented in the Pantex tracking system has been questioned by certain reviewers. In general, Pantex reported to OTA that it has responded to health and safety problems identified by the Tiger Team, through a variety of measures, including:

- increased personnel and budgeting;
- new policies to integrate health and safety into operations (environmental, safety, and health review system; computer tracking system);
- development of a site-specific RADCON manual and implementation plan to comply with new DOE requirements; and
- increased opportunities for communication between labor and management, and improved responsiveness of management to labor concerns (6).

A Progress Assessment of Pantex actions to address Tiger Team findings was conducted by DOE headquarters Office of Environment, Safety, and Health in March-April 1993. This assessment focused on four issues: 1) worker safety, 2) transportation of hazardous materials to and from Pantex, 3) high explosives, and 4) management of criticality safety. It concluded that Pantex has accomplished much in setting up new procedures at the facility, but the efforts are not well coordinated and implementation is lagging. One of the problems noted was a lack of independent DOE review of explosives safety. Lack of a program for transporting hazardous materials was also considered a problem. The lack of a formal program description for criticality safety was cited as an issue, although no workplace problems were identified in complying with procedures established by the nuclear weapons design laboratories. Regarding worker safety, M&H was considered to have an excellent program, and major improvements were identified in industrial hygiene; however, many safety problems were identified in construction projects managed by the U.S. Army Corps of Engineers (2).

DOE's own efforts at reviewing progress at Pantex have been constrained and fragmented in the past. Responsibility for internal oversight was divided between the Offices of Environment, Safety, and Health and Nuclear Safety (NS). Differing priorities of these two offices led to difficulties in coordinating facility reviews (25). Furthermore, beginning in September 1992, EH was required to give 60 days notice before inspecting a facility (71). This requirement delayed the Tiger Team Progress Assessment of Pantex (25).

Under the Clinton administration, Secretary of Energy Hazel R. O'Leary has made some changes. Nuclear Safety is again part of the Office of Environment, Safety, and Health, and the 60-day notice requirement for assessments has been rescinded. A number of other initiatives are under way to increase the independence and effectiveness of EH.[5] These initiatives and changes in headquarters oversight responsibility should be followed to see if they improve the ability of DOE to address important matters that cross the boundary between nuclear and nonnuclear issues, such as high explosives contaminated with radioisotopes, mixed waste, and the overall conduct of safety analyses and criticality safety.

A review of recent contractor performance evaluations can provide an understanding of environmental, safety, and health concerns at the dismantlement sites. DOE field offices routinely conduct performance evaluations of their M&O contractors (73). The Pantex evaluation for April-September 1992 reveals improvements as well as continuing deficiencies. The most notable improvements are in radiation safety and management attention to areas that can affect future operations at Pantex, including public affairs, environmental restoration, and nonradiological health and safety of workers. With regard to radiological protection, measures have been taken

[5] The May 5, 1993, Health and Safety Initiative includes developing a departmental safety and health policy; establishing the authority of the Assistant Secretary for Environment, Safety, and Health to force cessation of unsafe operations; rescinding the 60-day notice requirement for environmental, safety, and health assessments; developing a departmental "fatality policy" with strengthened investigation procedures; establishing employee-management health and safety committees for all Department sites; accelerating issuance of Price-Anderson nuclear safety rules; and initiating consultation with OSHA with the aim of establishing OSHA regulation of all Department sites (72).

to reduce employee exposures and waste handling requirements associated with dismantlement operations.

Improvements were made in radiation safety records, radiation worker training, and Radiological Assistance Team capabilities for emergency response operations. However, the evaluation report criticized other aspects of the emergency response program.

Some problems noted in the report illustrate how a lack of management attention to environmental restoration program activities affects worker health and safety. For example, a cleanup operation was delayed because of workers' not following the approved Health and Safety Plan, inadequate quality assurance for sampling, and improper placement of groundwater monitoring wells. These problems were attributed to inadequate oversight of a subcontractor. Also, potential noncompliances with RCRA were noted that could result in enforcement action from the Texas Water Commission.

DNFSB devoted considerable resources to Pantex in 1992, including several site visits and a full-time on-site representative (14,15). Site visits in March and August 1992 critiqued the status of safety analyses and criticality analysis at Pantex, and found that only a few Pantex facilities were covered by these programs. DNFSB also found that DOE's explosive safety guidelines did not give sufficient emphasis to nuclear material releases resulting from operational accidents that could occur in the disassembly cells. DNFSB identified concerns with the overall safety attitude during dismantlement operations, given the apparently wide latitude of disassembly workers to use their judgment instead of consulting a supervisor in the case of abnormal situations. DNFSB noted that disassembly technicians do not appear to be trained to question an operation that is not proceeding as expected. DNFSB also noted a need for criticality experts at Pantex to participate in the Nuclear Explosive Safety Study Group (NESSG) that reviews and approves all weapons assembly/disassembly procedures.

DNFSB staff has found both DOE and M&H to be responsive (40). One formal recommendation has been issued as a result of this review. It notes that there is now a discrepancy in nuclear safety requirements between facilities that produce and process fissile materials, and those such as Pantex that assemble, disassemble, and test nuclear weapons. DNFSB has recommended that DOE review its nuclear safety orders and directives, develop a plan to make nuclear safety assurances comparable at both types of facilities, and give priority to a site-wide compliance review at Pantex (13).

Y-12 PLANT

Numerous health and safety problems at Y-12 were identified by the DOE Tiger Team in its February 1990 report, including:

- slow progress in correcting procedural problems,
- slow progress in correcting training problems,
- radiation protection deficiencies,
- inability to attract and retain competent staff,
- numerous OSHA violations and a failure to follow basic safety guidance, and
- insufficient management attention to the environment, safety, and health (67)

The Tiger Team noted that Y-12 management paid insufficient attention to safety and health because it believed that new requirements were not cost-effective or necessary to protect worker health and safety, or that DOE did not really expect strict compliance with health and safety orders. Managers pointed to minimal DOE oversight, insufficient budgets from headquarters for health and safety, and the age and condition of Y-12 facilities, which make process alterations costly.

In February 1992, DOE returned to Y-12 for a selective review of environmental, safety, and health management systems and programs (84). In general, this assessment concluded that significant progress had been made in improving health and safety programs since the Tiger Team review, although some programs were still not in compliance with DOE requirements. Two key concerns were identified during this review: 1) insufficient formality and rigor in DOE and contractor oversight activities to ensure that environmental, safety, and health problems were identified and addressed effectively; and 2) inadequate use of available management tools by the contractor to correct environmental, safety, and health problems in this area.

The DOE Progress Assessment Team also identified some strengths at Y-12, including a Self-Assessment Program (lessons learned and alert system, division-level self-assessments, training curriculum), the ORR, and the Quality Assurance plan for the Nuclear Criticality Safety Department (84). However, the ORR process was found deficient by the DNFSB, as discussed below.

Y-12 has responded to Tiger Team criticisms by reorganizing and creating a new branch responsible for the environment, safety, and health. A new mission statement and strategic plan were as developed by the prime M&O contractor, Martin Marietta Energy Systems (MMES), in June 1991. Under the new organization and with the use of the ORR process that the Tiger Team identified as a strength at Y-12, environmental, safety, and health reviews are now integrated into the procedure for obtaining an internal license to begin a new operation. In the past, criticality safety was a major part of the review for new operations, but other health, safety, and environmental provisions were handled independently and often were not explicitly included in operating procedures. Training was the mechanism relied on to ensure workers were protected, but there were weaknesses in training too.

As of January 1993, Y-12 had established policies for integrating health and safety procedures into its operations. It is in the process of updating operating procedures, but this is time-consuming. In the dismantlement program, a project management team is established for each new weapons system that comes to Y-12 for dismantlement. Teams are composed of engineers, industrial hygienists, and health physicists, who define operating procedures, including health and safety. Updating procedures for dismantlement is expected to be completed in 1995-96. In the meantime, procedures are not considered to be "audit ready," and the health and safety aspects of operations depend on worker training (36).

In the performance evaluation of the Y-12 Plant (April to September 1992), DOE noted that—overall—safety and health programs were satisfactory, effective, and cost-efficient, consistent with DOE orders and applicable Federal, State, and local regulations providing protection for workers and the public (81). Individual health and safety programs were rated as good for industrial hygiene, nuclear facility safety, industrial safety/OSHA upgrades, nuclear criticality safety, and fire protection. Satisfactory ratings were also given for general safety and health, health physics/radiation protection, occupational medicine, and transportation safety.

However, implementation by direct line managers of programs, policies, and procedures to ensure protection of the environment, safety, and health, as well as quality, was rated as marginal. A marginal rating indicates a poorer performance than expected and has the potential to reduce the CPAF that could be received. Line management's commitment to a safe and healthy work environment was questioned because of the number, severity, and frequency of deficiencies in the implementation of safety and health programs, especially in radiation protection and contamination control.

For example, significant deficiencies identified included: 1) inadequate posting of respirator areas, which allowed personnel to enter areas without proper respiratory protection; 2) reentry of areas evacuated due to loss of ventilation, without respiratory protection or survey by health physics and industrial hygiene; and 3) criticality safety deficiencies associated with assessments that do not accurately reflect actual storage conditions, criticality safety procedures that do not incorporate actual requirements, and lack of review or approval of procedures (81).

On the other hand, there has been some notable progress in health and safety at Y-12 on the specific aspects of dismantlement of one weapons system (W-33). A new procedure was developed for dismantling a system that used existing chemical processes to remove protective coatings from parts, eliminating the generation of a new mixed RCRA waste (81).

To date, DNFSB has devoted fewer resources to Y-12 than to Pantex, and it conducted four site visits at Y-12 in 1992. Site visits in early 1993 identified continuing problems and lack of improvement in many areas. Staff concerns included compliance with orders/standards especially for radiological control, the ORR process, and training.

The problems with ORRs are notable because this area had been identified as a strength by the DOE Progress Assessment described above. The DNFSB criticized the ORR process at Y-12 in a March 1993 site visit. Inconsistencies with Board recommendation 92-6 were noted in several areas, including independence of senior members, scheduling, review criteria, and review team makeup (22).

There are other apparent discrepancies between DNFSB and internal DOE reviews. Despite satisfactory ratings in radiation protection in the April-September 1992 performance evaluation discussed above, DNFSB has identified continued deficiencies in radiological control practices at Y-12 vis-a-vis requirements found in DOE orders. The DNFSB requested a report from DOE evaluating the technical adequacy of radiological control practices compared with DOE and consensus standards, and a second report on plans to address longstanding problems of compliance with DOE orders (23).

Radiation Protection

Radioactive materials in nuclear warheads require measures to control health risks from occupational or environmental exposures for thousands of years. The radioactive half-lives (i.e., the time required for one-half of the material to undergo radioactive decay) of uranium-235 (the principal isotope in highly enriched uranium), plutonium-239, and plutonium-240 (the principal isotopes of plutonium in weapons) are approximately 700 million, 24 thousand, and 6.5 thousand years, respectively. The actual risk posed by a radioactive material is a combination of the material's half-life, the radiation emitted during decay, and its quantity. All three of these isotopes decay by the emission of an alpha particle accompanied by the emission of very weak x-rays. Uranium decay is also accompanied by the emission of some moderately energetic gamma-rays. There may also be some quantities of plutonium-241 present that decay to americium-241, which has x-ray and gamma-ray emissions accompanying its decay. The health effects of different types of ionizing radiation are described in box 3-D.

Alpha particles are easily stopped by materials as thin as a piece of paper, and x- or gamma-rays are shielded by the structural material surrounding intact warheads. Although there may be some exposure to penetrating external radiation from x-or gamma-rays during disassembly, the primary hazard arises from internal deposition of these isotopes via inhalation or ingestion, where the alpha particles are able to expose cells in internal organs, such as those lining the lung. Once inside the human body, alpha particles are much more damaging to surrounding tissue than other forms of penetrating radiation.

Harmful health effects are not likely to occur from being near plutonium unless one inhales or swallows it (88). Absorption through undamaged skin is limited, but plutonium can enter the body through wounds. Exposures are not likely from intact pits, which are usually clad with a protective coating. However, plutonium metal oxidizes rapidly in moist air or moist argon, forming a powdery surface coating. The corrosion or oxidation of plutonium does not always occur in a predictable manner and is affected by many variables, including the surrounding atmosphere, moisture content, and alloys or impurities present in the metallic plutonium (60). Uranium metal also oxidizes in air to form a coating that is easily removed as dust during handling of the metal (60). Plutonium oxide and fine particles of uranium metal can be a fire hazard because they are pyrophoric. Both the dust and the fires can result in inhalation hazards.

If ingested with food or water, most plutonium is poorly absorbed by the stomach and excreted. If inhaled, the amount remaining in the lungs depends on the particle size and form. Forms that dissolve easily may be absorbed and move to other parts of the body. Forms that dissolve less easily are often coughed up and possibly swallowed. Plutonium may remain in the lungs or move to the bones, liver, or other organs. It generally stays in the body for decades and continues to expose the surrounding tissues to radiation, which may eventually cause cancer. Cancer risks are naturally related to the level of exposure, but studies on the effects of low levels

Box 3-D—Health Effects of Ionizing Radiation

Radionuclides, such as the plutonium and uranium used in nuclear weapons, produce ionizing radiation that has the potential to cause biological damage. The material below represents a synthesis of information obtained by OTA (27,28,38,56,58,64).

Ionizing radiation is the transfer of energy through space in the form of either electromagnetic waves (e.g., x-rays, gamma-rays) or subatomic particles (e.g., alpha particles, neutrons) that are capable of separating electrons from their atomic or molecular orbits. Biological systems are highly structured and specific at the molecular level. The consequence of changes due to ionizing radiation is usually damaging to the function of the cell, tissue, or organ involved.

Cell damage from ionizing radiation may be reparable and cause no long-term problems, be imperfectly repaired and cause cell death, or be imperfectly repaired and lead to a modification in the cell, as discussed below. There is scientific controversy about the quantitative probability of adverse health effects due to low-dose exposure.

Reparable cell damage from ionizing radiation implies the existence of a dose threshold, or safe dosage, below which cells incur no damage from radiation exposure. This is a controversial issue in terms of low doses of radiation from gamma- and x-rays. There is solid scientific evidence that repair is far from perfect for high radiation doses and for doses from alpha particles.

Cell death may or may not adversely affect the functioning of tissues and organs. Most are unaffected by losses of even large numbers of cells. Radiation damage leading to cell death primarily affects nuclear DNA. Tissues that normally divide all the time (i.e., gastrointestinal tissue and bone marrow) are the most radiosensitive. These effects are *nonstochastic* or *deterministic:* the severity of the effect increases as the radiation dose increases.

Health effects on *modified cells* from ionizing radiation are best studied for cancer and birth defects. These effects are also *stochastic:* the likelihood that such an effect will occur increases as the radiation dose increases. Other effects have not been well studied (e.g., radiation that initiates inflammatory reactions in blood cells).

Ionizing radiation is classified into different categories for purposes of determining health effects and appropriate protection measures. *Penetrating or external radiation,* including x-rays, gamma-rays, and neutrons, can travel long distances and penetrate dense materials. To protect tissues from such radiation, shielding is used. *Particulate radiation* includes electrons, protons, alpha particles, neutrons, negative pi mesons (used for therapeutic radiation treatment), and heavy charged ions. Much particulate radiation is easily stopped by thin barriers, such as a piece of paper or skin, and can do biological damage only if deposited within tissue where it is referred to as an *internal emitter.* (Neutrons are an exception because they are also a form of penetrating radiation.)

Internal emitters are radionuclides that are taken into the body via inhalation, ingestion, or skin wounds. Once absorbed, the radioactive particles, depending on their half-life and decay chain, continue to emit radiation and potentially kill or alter cells for as long as they are lodged in the body. Internal emitters may be metabolized in ways that result in the radionuclide's becoming deposited in tissues for long periods.

The biological effects of internal emitters and penetrating radiation depend on the type and amount of radiation involved. Consequences of internal emitters also depend on the physical and chemical form of the radionuclide, the route of exposure, and metabolism. Small, soluble particles are generally the worst. Large particles are often removed from the body by natural processes, depending on the exposure pathway.

It is very difficult to determine the doses, effects, and risks associated with exposure to radiation. The biological effect of radiation correlates with dose but also depends on several other factors. Identical doses of different types of radiation, delivered in identical temporal patterns, produce different biological effects, even though the same amount of energy is delivered.

The type of radiation is especially important because of variations in the spatial distribution of the energy released. The "track" of a radioactive particle or wave is the set of all transfers of energy produced. Linear energy transfer (LET) is the amount of energy transferred and its spatial distribution per unit length of particle track (i.e, the amount of energy transferred per unit path traveled). Low-LET radiation (x-rays and gamma-rays) deposits energy throughout the cell in a diffuse pattern. High-LET radiation (alpha particles and neutrons) deposits energy in a much smaller volume, along fewer narrow, but dense, tracks. Biophysical data suggest that the ability of radiation tracks to produce damage in closely adjacent atoms and molecules determines the biological effect. Radiation types that produce dense patterns of ionization do more damage than those that spread the ionization over larger areas. Thus, alpha particles and neutrons will cause more damage than the same dose of gamma-rays, x-rays, and most beta radiation. A single track from a low-energy x-ray will produce about 1,000 ionizations along its length. High-LET radiation, such as an alpha particle, will produce about 1,000 times more ionization per unit distance, and large amounts of energy will be delivered to a very small molecular volume of tissue. High-LET radiation tracks can produce energy deposition patterns and concentrations that are improbable or impossible for low-LET tracks. Such patterns could translate into qualitatively different molecular biological effects.

SOURCE: Office of Technology Assessment, 1993.

of plutonium exposure are inconclusive. Animal studies link plutonium exposure to cancer and decreased ability to resist disease (88). It is not clear whether plutonium causes birth defects or reproductive problems.

The ingestion and inhalation hazards associated with weapons-grade uranium are considered the primary radiation hazard. Cancer risks due to enriched uranium are not known; however, there is some evidence that very long-term, low-level exposure to insoluble uranium causes increased risk of lung cancer (88). The chemical toxicity of uranium is a health concern, especially for soluble compounds that can cause kidney damage, but this is not the paramount concern when handling weapons-grade uranium metal, which is relatively insoluble (37).

Protecting workers and the public from exposure requires effective programs to prevent:

- criticality,
- unnecessary radiation exposures, and
- the unplanned release of radioactive material (60).

This can best be accomplished with well-trained workers functioning in well-designed, maintained, and monitored facilities with programs and policies that emphasize health and safety. DOE has recently upgraded its programs for radiation protection with the development of a new manual for radiation control. The new manual is a response to criticisms levied against DOE by numerous oversight bodies, including the Advisory Committee on Nuclear Facility Safety and the DNFSB (1,11,80).

Each DOE facility has been required to develop a compliance assessment and implementation plan (70). DOE progress in implementing this recommendation has been criticized repeatedly by DNFSB, although part of the problem may simply be associated with changing leadership of the Department (20). Major criticisms relate to DOE efforts to make organizational changes to ensure that excellence in radiation protection is a continuing priority and that the RADCON manual is not just a one-time effort to upgrade DOE programs (10). DNFSB is particularly concerned that appropriately trained people and adequate resources for radiation protection are available at all levels of DOE. Apparently, part of DNFSB's difficulty in evaluating the adequacy of DOE's response to its recommendation on radiation protection involves problems of access to the implementation plans of individual facilities.

OTA was able to obtain a draft copy of the implementation plan for Y-12. According to the plan, implementation of the DOE RADCON manual will require significant changes in Y-12 Plant facilities, equipment, and operational procedures (47).

Substantive changes are necessary for radiological assessment of the workplace, control of contamination, and employee awareness and understanding of radiological conditions and events. Y-12 does not anticipate being in substantive compliance until 1997. Although the plan lays out priorities and cost estimates, some significant costs are not included. Funding needed to address the requirements for containment and ventilation of existing equipment and facilities does not appear in the plan.

The DNFSB reviewed progress in implementing the RADCON manual at Y-12 in March-April 1993 (23). It noted a lack of technical justification for actions taken and for decisions made to defer or take no action. Major problem areas identified included contamination control (personnel monitoring, anticontamination clothing, break areas, personnel decontamination); training (facility-specific and core radiation worker); and occurrence reporting (skin contamination, clothing contamination, and compliance with a revised DOE order). Apparently, the DOE Oak Ridge Operations Office and its major contractor do not believe they have the resources to implement many of the mandatory requirements in the RADCON manual. DNFSB notes that Y-12 lags behind other DOE facilities in implementing the manual (23).

Worker Concerns About Health and Safety

The change in DOE culture to emphasize health, safety, and environmental concerns rather than production goals has been under way for several years. Internal and external oversight reports note accomplishments and the need for continuing improvement. Few of these reports, however, identify the views of workers and their sense of cultural change. OTA sought worker opinions on these changes through interviews by an anthropologist with labor representatives at DOE facilities (see appendix E).

Workers identified a number of areas as evidence that DOE and site managers are truly committed to improving health and operational safety, including more active DOE presence at individual facilities, greater contact with the union, and willingness to shut down a line if employees have a problem with a standard. Workers also cited more opportunities to resolve health and safety complaints, as well as greater responsiveness from management. Training improvements were noted.

However, several topics were identified as needing improvement, including unclear leadership and inconsistent enforcement; limited worker participation in developing improved procedures;

insufficient interaction between DOE and labor; budget limitations for implementing needed improvements in facilities; overburdened health and safety staff; continued fear of retribution for raising problems; deficiencies in health physics programs, policies, and instrumentation; and training problems.

A key question for many workers is whether all changes in health and safety practices are necessary, and whether increased oversight may at times contribute to the level of stress experienced by workers and thereby reduce overall job safety. Many of the changed procedures appear to be paperwork exercises that do not clarify the job of the worker. At the Pantex Plant, improvements are under way to make the standards more user friendly: for example, engineers visit the work floor, and workers participate in some procedure validation teams. Greater worker involvement in the development of procedures is desirable. Worker involvement has led to improvements that have reduced employee radiation exposures during dismantlement activities, and in one case, problems were encountered during a test of dismantlement procedures because of insufficient consultation with workers who had operational experience (73,81).

From a worker perspective, the change in emphasis from production only to concern for health, safety, and the environment can be aided significantly by improving opportunities for worker involvement in all levels of decisionmaking. The use of specially trained workers as health and safety representatives is viewed as a key to making worker concerns known to managers who have the authority to correct a problem. Worker empowerment is one of the new safety and health initiatives announced by the Secretary of Energy in May 1993. Guidance from EH to program offices and operations offices on establishing meaningful employee-management safety committees is planned to be implemented late in 1993.

Environmental Issues

Widespread environmental contamination with special nuclear materials during dismantlement activities is considered by DOE to be highly unlikely. Potential sources of such widespread contamination—however unlikely—include an inadvertent nuclear detonation, a criticality accident, or the scatter of nuclear materials from an explosion that does not cause a nuclear chain reaction. This conclusion is based on the record of handling and dismantling weapons in the past, analyses of risks conducted in Safety Analysis Reports (SARs), and criticality studies conducted by Los Alamos National Laboratory (LANL) and Lawrence Livermore National Laboratory (LLNL).

Although no major releases of special nuclear materials have been documented from Pantex dismantlement activities, because of the potential impact of major environmental contamination from dismantlement operations, there is concern that DOE should be far more aggressive in reducing the likelihood of such an occurrence. Reviewers of DOE operations have found fault with existing SARs, noting that they require updating (67,68), and have criticized some of the assumptions made about risk factors (e.g., the airplane crash scenario in the new SAR draft for double stacking plutonium pits in Pantex bunkers). DNFSB has formally expressed its reservations with Pantex SARs on two separate occasions and continues to review the SAR program at Pantex (15,18).

There is also concern by DNFSB about criticality safety at Pantex. The concern is primarily whether Pantex has sufficient rapid response capability to take care of any off-normal conditions associated with a warhead that may not be intact (41). DNFSB concerns stem from a lack of progress at Pantex on implementing DOE's own recommendations to improve criticality safety, which were identified in Defense Program's Technical Safety Appraisal (TSA) of Pantex in March 1992 (17). DNFSB required a report from DOE on progress in implementing TSA recommendations. The DOE response emphasizes that the design laboratories have not identified any credible potential for a criticality incident. However, DOE will implement its orders for SARs and criticality safety at Pantex, which were previously considered nonapplicable. Satisfactory implementation of these DOE orders, according to DNFSB, will result in significant upgrading of these areas. Pantex is also actively recruiting additional criticality staff. The ability to independently review design laboratory criticality analyses is delayed until the completion of a facility for classified computer operations scheduled for November 1993 (4).

A more recent review of criticality safety at Pantex was included as part of a Tiger Team Progress Assessment conducted by DOE's Office of Environment, Safety, and Health in March-April 1993. This assessment concluded that there were no workplace problems with criticality procedures (defined by the nuclear design laboratories); however, neither was there a good program description as called for by industry standards (2).

Experts at LANL conclude that the risks of a criticality accident at Pantex are extremely low, given that plutonium is present as a metal in well-defined shapes and stored in canisters in well-defined configurations (54). Criticality accidents are much more likely to occur during processing or management of liquid forms of plutonium or uranium as opposed to solids.

At Pantex, removing high explosives from intact weapons is one of the operations posing the highest risk of an accident that could scatter nuclear materials. However, these activities occur in special enclosures designed to contain explosions and releases. Once the nuclear materials have been separated from the explosives and detonators, the risks of explosion are significantly reduced.

The more likely sources of environmental contamination from dismantlement activities include:

- operational emissions,
- waste management practices,
- contamination from past practices, and
- contamination from old facilities.

The following discussion of environmental issues at Pantex and Y-12 offers examples from the sites most involved in dismantlement and is illustrative of the entire Weapons Complex.

PANTEX PLANT

One of the most likely sources of environmental contamination from projected warhead dismantlement operations at Pantex is the continued open burning of chemical explosives. This practice is used to separate some chemical explosives from other warhead materials that may be recycled or handled as waste, as well as to destroy some chemical explosives. Since 1943, Pantex has operated the Burning Grounds Facility for the combustion of high explosives or waste materials contaminated with high explosives, and for the disposal of scrap metal and other salvable weapons components (26,48). Liquid materials, including solvents, have also been burned there (52).

After burning, the remaining residue or ash from HE materials that are hazardous, and therefore subject to RCRA regulation, is collected and shipped off-site to a permitted facility for disposal (26,53). The scrap metal recovered after the burning of high-explosive materials and weapons components is collected for inspection and subsequently transported to another Pantex building for processing.

Emissions limits have been established for open burning operations. To reduce future emissions, RCRA permit conditions may become more restricted (45,63). Not all of the contaminants contained in materials burned at Pantex are covered by such permits. Radioactive emissions are subject to DOE orders within the facility boundary. Additional emissions from an increased rate of burning chemical explosives will also be an issue if certain warhead dismantlement activities are accelerated. Since 1988, however, DOE has taken action to keep the total amounts below annual permitted emission levels (29).

Another concern about continued open burning of chemical explosives is the release of toxic, volatile forms of fluorine. These are produced from burning chemical explosives that contain a fluorine-based plastic binder (3). Control measures have been implemented to reduce fluorine emissions, but the contamination potential may continue to exist within the Burning Grounds since fluorine limits apply to the entire facility rather than to individual burning units.

Pantex has traditionally handled disposal of some of the waste materials generated from dismantlement activities itself and shipped other materials off-site, either to different DOE facilities or for commercial disposal. Figure 3-2 shows the general disposition of waste materials from Pantex. Proper handling of wastes from dismantlement to prevent environmental contamination is now a Pantex responsibility that, at least for some materials, is subject to the scrutiny of State regulators.

Ongoing nuclear warhead dismantlement at Pantex generates parts that contain chemical explosives, special nuclear and hazardous materials, and classified components. From a regulatory perspective, dismantlement waste can be categorized as hazardous, radioactive, mixed (combined radioactive and hazardous), and nonhazardous. Nonhazardous waste includes scrap metal and trash. Scrap metal parts are collected and transported off-site by "outside bidders" for recycling (26), and trash is disposed of in landfills (75). Box 3-E describes how Pantex handles waste categories. Nuclear materials including plutonium, highly enriched uranium, and tritium are not classified as waste because they are currently considered assets by DOE (29).

Activities of M&H that involve the packaging and transportation of hazardous materials and wastes must comply with applicable DOE guidance. Recent inspections by DOE have raised concerns about M&H's inability to segregate wastes adequately, to provide clear shipping documentation, and to provide safe storage for radioactive mixed waste at certain facilities (50). M&H's methods of tracking stored explosives "... do not permit verification that explosive limits in storage areas are not exceeded," and written procedures for operations involving explosives lack review or expiration dates, and may not be reviewed as often as required. "Documentation seems to be the weakest point of M&H's explosive safety program" (49).

Environmental monitoring is being upgraded at Pantex in response to Tiger Team criticisms. The State of Texas and M&H are expanding environmental monitoring for both air and groundwater quality. However, there is still some controversy associated with new sampling strategies and quality assurance procedures (83).

One example of increased monitoring involves investigations to evaluate the potential contamination from past practices at the Burning Grounds. As part of the RCRA permit issued to the Pantex Plant in 1991, the Texas Water Commission requested DOE to investigate about 110 facilities, including the Burning Grounds, at which hazardous wastes were or are currently being managed, stored, or disposed. After a recent visual inspection of sites (e.g., percolation/evaporation pit, burn pads, and landfills) at the Burning Grounds (87), DOE plans to submit a RCRA Facility Investigation report to State regulators in July 1994 (57,66).

The start of this investigation has been delayed by the need for DOE to obtain permit modifications for mixed waste storage capacity to handle any waste from drilling mud (57). Depending on the extent of environmental impacts identified in the final report, the Texas Water Commission will recommend specific measures for DOE to solve any contamination problems (79). Pantex officials also plan to complete by September 1993 a report on closing certain operations at the Burning Grounds in accordance with National Environmental Policy Act requirements (75).

Box 3-E—Waste Disposal at Pantex

Classified Waste

Classified parts must be declassified prior to treatment, recycling, or ultimate disposal to avoid disclosure of restricted data (8). Such parts are processed to declassify them at Pantex and then transported for recycling or ultimate disposal. Parts that cannot be declassified are transported to other DOE sites for disposal (e.g., the Nevada Test Site). The increasing cost of off-site management of such classified components has led to a plan to build a treatment, processing, and declassification facility at Pantex. Before its construction, a classified hazardous waste staging facility will be used there.

RCRA-Regulated Hazardous Waste

Pantex warhead dismantlement activities generate RCRA-regulated hazardous materials. The largest RCRA-regulated waste streams include chemical explosives and chemical explosives-containing parts. Chemical explosives and parts are burned in the open at the Burning Grounds, and the ash is collected and disposed at an approved off-site RCRA Hazardous Waste Disposal Facility. Other contaminated materials are burned, and RCRA hazardous residues are accumulated at a permitted storage area. Mixed waste or radioactive waste residues are not expected from Burning Ground operations.

Radioactive and Mixed Waste

With the exception of waste known to contain hazardous materials regulated under RCRA, all radioactive waste is transported to the Nevada Test Site for disposal. Low-level radioactive waste generated at Pantex production and disassembly areas is generally in solid form. It is collected, packaged, labeled, and moved to storage bunkers to await off-site transportation. Mixed wastes are currently being stored on-site in permitted areas, pending development of treatment options. Little information is available about current storage capacity limitations, the management of mixed waste, the potential implications of increased weapons dismantlement, and costs associated with off-site treatment and disposal of mixed waste.

SOURCE: U.S. Department of Energy, 1993.

Most of DOE's Weapons Complex facilities, including those used for warhead dismantlement, were built more than 30 years ago. The facilities in use today have generally been upgraded, but some still employ processes and technologies that are inefficient and create large amounts of waste. Some are also burdened by the safety and environmental legacy of past operations. Identification of obsolete or inefficient DOE facilities and technologies will be critical for ensuring long-term safe management of the materials from warhead dismantlement. Determining which facilities must be upgraded or replaced will be a challenge. DOE has indicated that "piecemeal improvements have proven inadequate" (69).

The Burning Grounds at Pantex constitute a specific example of an obsolete and potentially dangerous facility used in the current warhead dismantlement program. Since 1943 it has been operated for incineration of high explosives, salvable weapons components, and materials contaminated with high explosives (including foams, plastics, metals, solvents, and trash) (48,63). Although the facility is permitted to operate through 2001, there are serious concerns about future and continued operation. Several studies are under way to evaluate current conditions and identify possible upgrades. However, it seems that extensive and costly modifications may not ensure long-term environmentally safe management of high explosives, and compliance problems could result.

The dilapidated condition of the Pantex underground sewer system is also a problem. In

Explosion-resistant warhead disassembly cells at Pantex (Gravel Gerties).

addition to cracked or broken pipes, it suffers from excessive sediment deposits and blockages. The Pantex Tiger Team reported that the sewer system is obsolete and inadequate because it was originally constructed during World War II to accommodate the extensive flows associated with ammunition production, as opposed to the much smaller flows generated by current activities. Work is under way to replace some pipes; however, additional funding is required to support the program (33).

Y-12 PLANT

The Y-12 Plant at Oak Ridge, Tennessee is an industrial center that processed uranium and other materials for weapons production. Present activities also include processing and storage of uranium coming from warhead dismantlement. Like Pantex dismantlement activities, Y-12 processes generate classified, radioactive, hazardous, and nonhazardous wastes. Examples include machine turnings and metal fines; uranium- and beryllium-contaminated trash; waste solutions from metal plating; liquid waste and sludge generated by processing operations or waste treatment activities; and waste oils and solvents derived from machining and cleaning activities (86). These wastes are handled at Y-12 or transported off-site for treatment and final disposition.

The list of chemical substances—both radioactive and nonradioactive—used at Y-12 is extensive. This complicates monitoring of possible environmental releases into the air or water of contaminants from disposal operations at the plant. Some releases of toxic materials have been monitored. The processing and waste treatment operations at Y-12 discharged 884 pounds of uranium in 1990 to nearby surface water bodies and 46 pounds into the atmosphere in compliance with State permit conditions. Even though a new treatment facility has replaced depleted uranium land disposal, contamination from past land burial practices remains unresolved. Another source of uranium emission is accidental ignition of enriched uranium chips or saw fines stored in nonprocess areas at uranium production buildings. For example, in December 1985 a basket containing nearly 8 pounds of enriched uranium chips caught fire, leading to the release of enriched uranium.

Operations at Y-12 have generated large quantities of low-level radioactive solid waste. In 1990, DOE reported generating 4.3 million pounds of low-level radioactive waste, or about 22 percent of the total volume of all contaminated waste (including hazardous waste) produced at Y-12 that year (86). Uranium is the most common radioactive waste material, but other radioactive contaminants were reported, including fission products, thorium, and transuranic elements. Mixed waste storage is a problem at Y-12 because of the limited availability of treatment technologies, storage capacity limitations, and an increasing inventory of mixed waste from site cleanup activities.

The generation and management of hazardous waste at Y-12 are regulated by the Tennessee Department of Conservation under RCRA authority. Because treatment and storage of this waste are conducted at several buildings under State or Federal permit, waste inventories are recorded in several different databases. As a result, making a quantitative estimate of plant-wide waste is difficult. Similarly, predicting future generation rates, or estimating current compliance status at Y-12, is problematic because of the lack of consistent databases. Plans to adopt a comprehensive tracking system are under development.

RCRA compliance has been the most challenging environmental requirement for Y-12.[6] According to the Performance Evaluation Report for April-September 1992, Y-12 received outstanding or good ratings for all other environmental laws and a satisfactory rating for RCRA. The internal RCRA inspection program conducted by the M&O contractor has had difficulties obtaining adequate data and ensuring that RCRA is applied consistently (81).

Although operations at Y-12 have recently been reduced, it still faces several environmental problems based on past practices, including human and environmental health impacts of past radioactive emissions, adequate retrofitting of old facilities or building of new environmentally safe facilities, and provision of appropriate treatment technologies and storage for current waste inventories. There are growing numbers of interim and long-term cleanup projects under way at Y-12 to address environmental contamination from past activities. These remediation activities will have to be conducted concurrently with warhead dismantlement.

Ongoing environmental problems could impact future warhead dismantlement activities at Y-12. For example, ambient air levels of radioactive, inorganic, and organic materials are an ongoing problem. The 1990 Tiger Team assessment noted a lack of emission control and monitoring devices at several nonprocessing areas in uranium processing buildings. It also noted a lack of adequate documentation and control of radiological conditions at the Y-12 Plant; an incomplete survey of operational and radiological areas that are potential sources of worker exposure to radiation; and limited documentation to verify compliance with air emission regulations. There are no non-DOE Federal or State standards specifically for atmospheric emissions of uranium, although total radionuclide levels are regulated through the Clean Air Act.[7] Future warhead dismantlement operations could affect ambient levels of radionuclides at Oak Ridge. Recent reports have noted a very slight increase in local airborne concentrations of radionuclides (86).

In addition, beryllium, processed at Y-12 in both metallic and oxide powder form, is a potential health hazard and environmental contaminant. Although ambient air levels of beryllium are below State regulatory standards, the Y-12 Tiger Team recommended that a new beryllium monitoring system be installed because the existing system was outdated (67). In addition to uranium and beryllium at Y-12, the release of volatile organic compounds, such as chlorinated solvents used as degreasers (mostly perchloroethylene), cleaners, and machining coolants, has

[6] Until the mid-1980s, DOE maintained that the Atomic Energy Act exempted the hazardous portion of mixed waste from regulation under RCRA. A Federal court rejected DOE's position. See *Legal Environmental Assistance Foundation* v. *Hodel*, 586 F. Supp. 1163 (E.D. Tenn. 1984).

[7] U.S.C.A. (1251-1376 (West 1983, Supp. 1990)).

been reported. Substantial emission reductions have been accomplished at the plant since 1985, in part due to substituting less hazardous materials in certain Y-12 operations, as well as overall cutbacks in production (46).

CONCLUSIONS

Nuclear warhead dismantlement activities are ongoing at several sites in the DOE Nuclear Weapons Complex. The rate of dismantlement at the Pantex Plant has been reasonably constant over the past few years and is expected to increase only moderately in the future. The current rate of about 1,400 warheads per year has not put undue strains on the physical facilities, but material flows and waste streams have changed the focus of the operating plants. A key issue is whether or not the new systems now in place and under development to address environmental, safety, and health issues are sufficient to manage the dismantlement program. Clearly, there have been problems with past practices at dismantlement facilities. Many of these problems have been publicized in recent years, and efforts have been made to improve the situation. Improvements have been achieved in some areas, yet problems continue to be documented by both internal and external oversight activities. Several years will be required to integrate new policies and procedures into some operations, and DOE will need to continually review whether adequate progress is being made.

Resource requirements to implement environmental, safety, and health programs will be demanding and will continue for many years. Capital expenditures are needed to establish environmental monitoring programs and to improve plant conditions for workers. It will be important to ensure that these improvements are well planned and efficiently implemented.

The repeated criticism of lack of line management attention to environmental, safety, and health issues has been particularly troublesome. Improvements in programs for health physics,

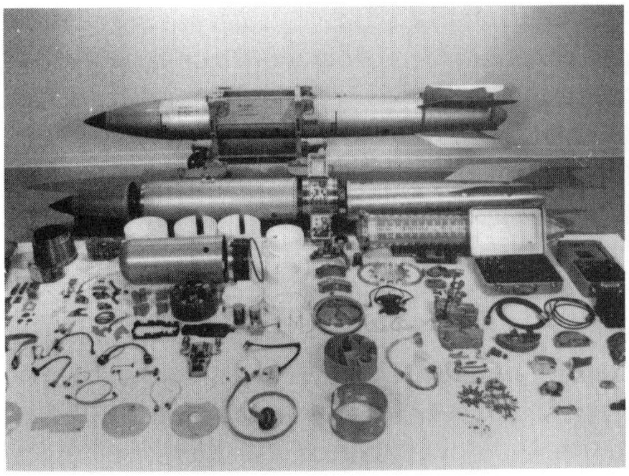

U.S. DEPARTMENT OF ENERGY

Many of the more than 6,000 parts contained in a B-61 nuclear bomb are displayed here, along with an intact weapon and its four major subassemblies.

industrial hygiene, and occupational safety are commendable, but it is unlikely that these programs will ever have enough resources and authority to get the job done without the additional support of the DOE Office of Defense Programs, which needs to devote more resources and attention to environmental, safety, and health issues. Providing support for greater worker involvement in the development of new procedures, more opportunities for worker control of health and safety during operations with the expansion of the Health and Safety Representative program, and additional opportunities for workers to communicate with DOE, not just with M&Os, may help DP become more proactive on these issues.

Nuclear weapons production history and recent developments in DOE operations have demonstrated that although some change is evident, continued scrutiny by outside parties will be necessary to ensure that progress continues in improving protection of the environment, safety, and health during dismantlement activities. Routine review or approval of plans by DNFSB prior to initiating new weapons dismantlement programs is desirable and may require legislative action to broaden the statutory scope of the Board. Additional resources to expand external Federal or State oversight into areas, such as health issues and environment monitoring, which are currently not the primary concern of DNFSB, are also desirable. Oversight by outside parties can add to the credibility of the review, given that there have been examples in which DNFSB has been more critical than DOE internal reviews.

Public credibility has the potential to become a major issue affecting the progress of dismantlement. Although DOE has acknowledged the importance of this in its evaluation of the Pantex program, no significant changes have been made in methods to communicate with the public, to understand public concerns, or to involve the public in decisions. Major attention must be given to this issue to ensure that public participation programs are effective. Not only must DOE and its contractors involve the public in their activities and decisions, but outside oversight bodies such as the DNFSB must also provide expanded opportunities for public participation.

CHAPTER 3 REFERENCES

1. Advisory Committee on Nuclear Facility Safety to the U.S. Department of Energy, "Final Report on DOE Nuclear Facilities," November 1991.
2. Barber, Bob, U.S. Department of Energy, Office of Environment, Safety, and Health, personal communication, Mar. 8, 1993.
3. "Basis for Fluorine and Other Pollutant Calculations," Appendix 4 of Engineering Instruction No. HE 92-019 (Lifting of Fluorine Burn Restrictions for TABT-Based Explosives), Mar. 2, 1992.
4. Beckner, E., Defense Nuclear Facilities Safety Board, letter to Conway, DNFSB Document No. 93:1058, Mar. 3, 1992.
5. Boardman, K., Weapons Programs Division, DOE Albuquerque Operations Office, briefing of Office of Technology Assessment staff on "WPD Perspective on Dismantlement," May 21, 1992.
6. Burr, John, Office of Environment, Safety, and Health, Batelle-Pantex, personal communication, April 1992.
7. Caskey, W., Sandia National Laboratories, briefing on "Project 'Stage Right'," May 20, 1992.
8. Chavez, J.J., U.S. Department of Energy, Albuquerque Operations Office, Office of Chief Counsel, memorandum to Michael J. Eckart, Weapons Programs Division, regarding "Transporation of Classified Weapons Parts to DOE Facilities for Declassification and Disposal," Sept. 12, 1991.
9. Clements, T., Southeast Nuclear Campaigner, Greenpeace, Washington, DC, personal communication, June 1993.
10. Davis, J., Defense Nuclear Facilities Safety Board, personal communication, Apr. 27, 1993.
11. Defense Nuclear Facilities Safety Board, Recommendation No. 91-6, "Radiation Protection for Workers and the General Public at DOE Defense Nuclear Facilities," Dec. 19, 1991.
12. Defense Nuclear Facilities Safety Board, *Annual Report to Congress* (Washington, DC: Defense Nuclear Facilities Safety Board, February 1992).
13. Defense Nuclear Facilities Safety Board, Fiori letter to Conway, Chairman, DNFSB Document No. 92:3447, June 17, 1992.
14. Defense Nuclear Facilities Safety Board, memorandum: Pantex Trip Report: July 7-9, 1992, DNFSB Document No. 92:4110, attachment to letter from Conway to Admiral Watkins, Secretary of Energy, July 30, 1992.
15. Defense Nuclear Facilities Safety Board, memorandum: Pantex Trip Report: August 11-14, 1992, DNFSB Document No. 92:4803, attachment to letter from Conway to Admiral Watkins, Secretary of Energy, Sept. 11, 1992.
16. Defense Nuclear Facilities Safety Board, Conway letter to Admiral Watkins, Secretary of Energy, DNFSB Document No. 93:0607, Dec. 22, 1992.
17. Defense Nuclear Facilities Safety Board, Conway letter to Admiral Watkins, Secretary of Energy, regarding "Nuclear Criticality Safety at the Pantex Plant," DNFSB Document No. 93:004, Dec. 31, 1992.
18. Defense Nuclear Facilities Safety Board, Conway letter to Stuntz regarding "Observations from a Trip to Pantex to Evaluate the W-79 DOE Operational Readiness Evaluation (ORE)," DNFSB Document No. 93:0369, Jan. 21, 1993.
19. Defense Nuclear Facilities Safety Board, Recommendation 93-1, "Standards Utilization in Defense Nuclear Facilities," Jan. 21, 1993.

20. Defense Nuclear Facilities Safety Board, Conway letter to H. O'Leary, Secretary of Energy, Mar. 23, 1993.
21. Defense Nuclear Facilities Safety Board, *Annual Report to Congress* (Washington, DC: Defense Nuclear Facilities Safety Board, April 1993).
22. Defense Nuclear Facilities Safety Board, Conway letter to Beckner, DNFSB Document No. 93:2022, Apr. 21, 1993.
23. Defense Nuclear Facilities Safety Board, Conway letter to H. O'Leary, Secretary of Energy, DNFSB Document No. 93:3290, June 8, 1993.
24. *Energy Daily,* vol. 21, No. 68, Apr. 12, 1993.
25. Fitzgerald, J., U.S. Department of Energy, Office of Environment, Safety, and Health, personal communication, Mar. 8, 1993.
26. Gilmore, S.K., Programs Management, Self-Assessment Section, ''Facility Operations Division: Waste Streams,'' February 1992.
27. Gofman, John, *Radiation Induced Cancer from Low-Dose Exposure: An Independent Analysis* (San Francisco, CA: Committee for Nuclear Responsibility, 1990).
28. Goodhead, D., ''Spatial and Temporal Distribution of Energy,'' *Health Physics*, vol. 55, No. 2, August 1988, pp. 231-240.
29. Goodwin, T., Weapons Development and Acquisitions Division, Defense Programs, U.S. Department of Energy, personal communication, Jan. 21, 1993 and Feb. 10, 1993.
30. Googin, J., Senior Staff Consultant, Martin Marietta Energy Systems, Oak Ridge, personal communication, June 1993.
31. Grant, J., Lt. Col., U.S. Department of Energy, Project Leader for Nuclear Weapons Retirement and Disposal, briefing to Office of Technology Assessment, May 20, 1992.
32. Guidice, S., Office of Operations and Weapons, Albuquerque Field Office, U.S. Department of Energy, briefing to Office of Technology Assessment, May 20, 1992.
33. Hallein, N., U.S. Department of Energy, personal communication, May 25, 1993.
34. Harrison, J., Weapons Project Group, Sandia National Laboratories, briefing to Office of Technology Assessment, ''Sandia's Role in Dismantlement Relative to Environmental Safety,'' May 20, 1992.
35. Hernandez, T. Sandia National Laboratories, Advanced Fuzing Department 5166, briefing to Office of Technology Assessment, ''Dismantlement System Studies and Model Developments,'' May 20, 1992.
36. Hoag, D., Office of Environment, Health and Safety, Oak Ridge Y-12 Plant, personal communication, January 1993.
37. Idaho National Engineering Laboratory, ''Health Physics Manual of Good Practices for Uranium Facilities,'' DE88-013620, June 1988.
38. International Commission on Radiation Protection, *1990 Recommendations of the International Commission on Radiation Protection*, ICRP Publication 60 (Oxford, England: Pergamon Press, 1990).
39. Jones, A., Intelligent Systems and Robotics Department, Sandia National Laboratories, briefing to Office of Technology Assessment, ''Sandia Robotics Capabilities,'' May 20, 1992.
40. Krahn, Steve, Defense Nuclear Facilities Safety Board, personal communication, Feb. 2, 1993.
41. Krahn, Steve, Defense Nuclear Facilities Safety Board, personal communication, June 6, 1993.
42. Lombardo, K., Albuquerque Field Office, Weapons Programs Division, U.S. Department of Energy, letter to Steve Lipmann, Environmental Research and Analysis, Oct. 1, 1992.
43. Longenbaugh, C., Albuquerque Field Office, Weapons Programs Division, U.S. Department of Energy, letter to Steve Lipmann, Environmental Research and Analysis, Sept. 29, 1992.
44. Longmire, Paul, Dismantlement Program Manager, Sandia National Laboratories, Oct. 15, 1992.
45. Lowerre, R., Henby, Kelly & Lowerre, letter to Beverly Gattis, State of Texas Alliance for Resources, Dec. 30, 1991.
46. Lynch, R., Director, Y-12 Facility Management Division (DP-43), U.S. Department of Energy, personal communication, May 18, 1993.
47. Martin Marietta Energy Systems, Inc., ''Oak Ridge Y-12 Plant, Compliance Assessment and Implementation Plan for the DOE Radiological Control Manual,'' Y/DQ-31, Revision O, Sept. 21, 1992.
48. Mason & Hanger-Silas Mason Co., Inc., Environmental Protection Department, Environmental Monitoring Session, ''Pantex Plant Site Environ-

mental Report for Calendar Year 1990," MHSMP-91-06, July 1991.

49. Mason & Hanger-Silas Mason Co., Inc., "Explosives Safety Appraisal, Mason & Hanger-Silas Mason Co., Inc., Pantex Plant, January 28-February 6, 1992," undated.
50. Mason & Hanger-Silas Mason Co., Inc., "Packaging and Transportation of Hazardous Materials Safety Appraisal, Mason & Hanger-Silas Mason Co., Inc., Pantex Plant, Amarillo, Texas, January 28-February 6, 1992," undated.
51. Mason & Hanger-Sila Mason Co., Inc., "Pantex Site Maintenance Plan, Vol. IV: Structures, Systems, and Components Description (SSCD) for Pantex Facilities" (Issue 1: Non-Reaction Nuclear Facilities), MNL-FO-1103 PROG, Issue No. 1, Mar. 2, 1992.
52. Mason & Hanger-Silas Mason Co., Inc., "Burning Grounds," fact sheet provided as part of briefing package distributed to the Public Information Coordinating Group, May 20, 1992.
53. Mathews, W., Mason & Hanger-Silas Mason Co., Inc., "Waste Management—Pantex Waste Buster," briefing package, circa 1992.
54. McLaughlin, T.P., Group Leader, Nuclear Criticality Safety, Los Alamos National Laboratory, personal communication, Mar. 10, 1993.
55. Mitchell, B., Nuclear Safety Campaign, Seattle, WA, personal communication, June 1993.
56. Modan, B., "Cancer and Leukemia Risks After Low-Level Radiation—Controversy, Facts and Future," *Medical Oncology and Tumor Pharmacotherapy*, vol. 4, No. 3/4, 1987, pp. 151-161.
57. Muck, C., Texas Water Commission, personal communication, May 18, 1993.
58. National Research Council, *Biological Effects of Ionizing Radiation* (Washington, DC: National Academy Press, 1989).
59. National Research Council, *The Nuclear Weapons Complex, Management for Health, Safety, and the Environment* (Washington, DC: National Academy Press, 1989).
60. Pacific Northwest Laboratory, "Health Physics Manual of Good Practices for Plutonium Facilities," DE88-013607, May 1988.
61. Soden, C., Abuquerque Field Office, Environmental Protection Division, U.S. Department of Energy, briefing to Office of Technology Assessment, "Environmental Protection, Waste Management, Air Quality, NEPA, and Compliance," May 21, 1992.
62. Tasking Memorandum from Rear Admiral Ellis, Deputy Assistant Secretary for Military Applications, to the national laboratories and to the Department of Energy Office of Operations and Weapons.
63. Texas Water Commission, "Hazardous Waste Permit 50284," Pantex Plant, Amarillo, undated.
64. United Nations Scientific Committee on the Effects of Atomic Radiation, *Sources, Effects and Risks of Ionizing Radiation—UNSCEAR 1988 Report* (New York: United Nations, 1988).
65. U.S. Congress, Office of Technology Assessment, *Hazards Ahead: Managing Cleanup Worker Health and Safety at the Nuclear Weapons Complex*, OTA-BP-O-85 (Washington, DC: U.S. Government Printing Office, February 1993).
66. U.S. Department of Energy, "Environmental Restoration Program," briefing package, GA92-824/LE1-LE8, undated, p. LE6.
67. U.S. Department of Energy, *Tiger Team Assessment of the Oak Ridge Y-12 Plant, Oak Ridge, Tennessee*, DOE/EH-0108 (Washington, DC: U.S. Department of Energy, February 1990).
68. U.S. Department of Energy, *Tiger Team Assessment of the Pantex Plant, Amarillo, Texas*, DOE/EH-0130 (Washington, DC: U.S. Department of Energy, February 1990).
69. U.S. Department of Energy, *Nuclear Weapons Complex Reconfiguration Study*, DOE/DP-0083 (Washington, DC: U.S. Department of Energy, January 1991).
70. U.S. Department of Energy, Notice DOE N 5480.11, Radiation Protection for Occupational Workers, June 17, 1992.
71. U.S. Department of Energy, Notice DOE N 2321.1, Coordinating and Scheduling Reviews, Assessments, and Other Oversight Activities at Department of Energy Field Offices/Elements, Sept. 28, 1992.
72. U.S. Department of Energy, Safety and Health Initiative, DOE News, R-93-085, May 1993.
73. U.S. Department of Energy, Albuquerque Operations Office, Award Fee Performance Evaluation Report for the period: April 1, 1992-September 30, 1992, Contract DE-AC04-91AL65030, Mason

& Hanger-Silas Mason Co., Inc., Pantex Plant, undated.

74. U.S. Department of Energy, Albuquerque Operations Office, "FY 1994 Budget Request, Construction Project Data Sheet, Environmental Restoration and Waste Management, Waste Management-Treatment Operations," March 1992.
75. U.S. Department of Energy, Albuquerque Operations Office, "Environmental Restoration and Waste Management Five Year Plan Activity Data Sheet FY 94-98," ALPX-3102, May 28, 1992.
76. U.S. Department of Energy, Albuquerque Operations Office, "Environmental Restoration and Waste Management Five Year Plan Activity Data Sheet FY 94-98," ALPX-3101, ALPX-3103, ALPX-3104, and ALPX-3105, May 28, 1992.
77. U.S. Department of Energy, Albuquerque Operations Office, "Occurrence Report ALO AO-MHSM-PANTEX-1992-0028," July 10, 1992.
78. U.S. Department of Energy, Albuquerque Operations Office, "Occurrence Report ALO-AO-MHSM-PANTEX-1996-0068, Unexpected Cracking of Weapon Component," July 26, 1993.
79. U.S. Department of Energy, Amarillo Area Office, "Pantex Plant Evironmental Restoration Program—Fact Sheet," undated.
80. U.S. Department of Energy, Assistant Secretary for Environment, Safety and Health, "Radiological Control Manual," DOE N 5480.6, June 1992.
81. U.S. Department of Energy, Oak Ridge Operations Office, "Performance Evaluation Committee Report for Martin Marietta Energy Systems, Inc.," Weapons Business Unit, Contract No. DE-AC05-84OR21400, Evaluation Period: Apr. 1, 1992-Sept. 30, 1992.
82. U.S. Department of Energy, Office of Defense Programs, Office of Inspections, "Technical Safety Appraisal of the Pantex Plant, Amarillo, Texas," March 1992.
83. U.S. Department of Energy, Office of Defense Programs, information provided to Office of Technology Assessment during a site visit to Pantex Plant, Amarillo, TX, February 1993.
84. U.S. Department of Energy, Office of Environment, Safety, and Health, *Progress Assessment of the Oak Ridge Y-12 Plant, Oak Ridge, Tennessee,* DOE/EH-0265 (Washington, DC: U.S. Department of Energy, February 1992).
85. U.S. Department of Energy, Office of Environment, Safety, and Health, briefing to Office of Technology Assessment, Dec. 8, 1992.
86. U.S. Department of Energy, Office of Environmental, Safety, and Health Compliance, *Oak Ridge Reservation Environmental Report for 1990 —Volume 1: Narrative, Summary, and Conclusions,* EH/ESH-18/VI (Oak Ridge, TN: U.S. Department of Energy, September 1991).
87. U.S. Department of Energy, Pantex Plant, Amarillo, TX, "Final Community Relations Plan for RCRA Investigations," Appendix D, May 1991.
88. U.S. Department of Health and Human Services, Public Health Service, Agency for Toxic Substances and Disease Registry, "Toxicological Profile for Uranium," TP-90-21, prepared by Syracuse Research Corporation under subcontract to Clement International Corporation, Contract No. 205-88-0608, December 1990.
89. Watkins, James, Secretary, U.S. Department of Energy, "Ten-Point Initiative—Remarks by James D. Watkins, June 27, 1989," reprinted in "Evaluation of the U.S. Department of Energy's Occupational Safety and Health Program for Its Government-Owned Contractor-Operated Facilities," Occupational Safety and Health Administration, December 1990, pp. B-1 to 13.
90. *Weapons Complex Monitor,* "The Supreme Court Has Refused to Overturn a Lower Court's Ruling," vol. 4, No. 26, June 9, 1993, p. 29.
91. West, G.T., Mason & Hanger-Silas Mason Co., Inc., Pantex Plant, Amarillo, TX, "United States Nuclear Warhead Assembly Facilities (1945-1990)," March 1991.
92. White, D.G., U.S. Department of Energy, Albuquerque Operations Office, Amarillo Area Office, Administrative Management Branch, memorandum on "Dismantlement of Nuclear Weapons," to Frank A. Baca, Division Director, undated.
93. Woods, R.O. "The Role of Robotics in Reducing the Nuclear Weapons Inventory," written for Office of Technology Assessment, Aug. 28, 1992.

Future Disposition of Nuclear Materials From Dismantled Weapons 4

After nuclear weapons are taken apart, the nuclear materials that contained such massive destructive power remain. Two principal materials—plutonium and highly enriched uranium (HEU)—are the most problematic. Together or separately they could be made into new weapons. Thus there are serious concerns about keeping these materials both safely contained and securely guarded. This chapter analyzes the management (disposition) of these materials; current plans for their storage, further use, processing, or disposal; studies that are addressing various technical approaches for disposition; and policies that affect these decisions. The discussion focuses on materials from U.S. warheads, although the technology for storage and disposition can have international application. Chapter 6 discusses possible application to Russian weapons materials.

Both plutonium and HEU can be used to make nuclear warheads, either in combination or alone. Although modern nuclear warheads commonly use both materials, the "Fat Man" and "Little Boy" U.S. atomic bombs used in 1945 contained exclusively plutonium or HEU, respectively. Nevertheless, the ease of making a simple bomb from each material is quite different. A HEU bomb would be easier to design than one using plutonium, and would offer a higher confidence of working

Point

"We paid dearly for [weapons plutonium] in terms of dollars and the environment—let's get something back!"

DOE weapons design laboratory reviewer of OTA report

Counterpoint

"The DOE presumption of plutonium as an asset . . . is a significant policy issue that needs to be decided by Congress, not a group of career civil servants and cold warriors within DOE."

Citizen group reviewer of OTA report

without being tested than a similar plutonium-based bomb (36).[1] With the first U.S. nuclear bombs, only the plutonium-based design was tested before use.

This chapter discusses a variety of ideas for storing, utilizing, processing, and disposing of the plutonium and HEU recovered from dismantled warheads. Some consider HEU to pose a much simpler problem because there is an existing market for uranium fuel. Conversion of surplus HEU into conventional low-enriched uranium (LEU) fuel for use in existing nuclear reactors is technically straightforward. An existing U.S. reactor could use fuel from diluted weapons-grade material just as easily as fuel from conventional sources. On the other hand, it will take decades to convert large quantities of HEU in this manner; during that time the HEU will have to be stored, and will present a continuous risk of proliferation and diversion. In fact, if proliferation resistance were the *only* criterion by which to judge disposition options, one might actually consider options such as glassification of HEU with high-level waste—an option that is being considered seriously only for plutonium.

Plutonium may present a more difficult disposition problem. No civilian power reactors in the United States currently use plutonium for fuel, and although its use is technically feasible, the political and regulatory obstacles may be enormous. In addition, the United States chose to abandon the use of plutonium fuel in commercial reactors nearly two decades ago for political, security, and economic reasons, and it would be difficult to resurrect this effort. Therefore, it is likely that a greater number of possible options will have to be examined for plutonium than for HEU. In any event, considerably more literature is available about the disposition of plutonium than about HEU. The disparity is reflected in this report: the section analyzing plutonium options is considerably longer than that devoted to HEU.

Both plutonium and HEU have extremely long half-lives (24,000 years for plutonium-239 and orders of magnitude longer for the isotopes of uranium in HEU). They will, therefore, need to be contained or isolated for long periods to prevent environmental contamination or possible human intrusion and exposure. Both of these materials pose health risks, as described in chapter 3. Plutonium is especially toxic in minute quantities if inhaled or ingested.[2]

The amount of plutonium and HEU from retired weapons is growing, as is the need to do something with it. This chapter is about "disposition"[3] of this material, specifically the spectrum of possibilities about what to do with it beyond weaponry: destroy some portions if technically feasible and practical; dispose of some as waste if technology and national policies permit; or utilize some to produce civilian energy, if security is adequate and if technology and economics prove sound.

A few hundred tons of weapons-grade plutonium and more than a thousand tons of HEU (exact numbers are classified) exist in the world today—as either intact warheads; forms ready to be made into warheads, pits, and other components removed from retired weapons; or residues from the past manufacture of plutonium for weapons (75-77). The United States and Russia have by far the majority of these materials. Both plutonium and uranium are also found in various forms and quantities in the nuclear industry

---

[1] However, HEU is much harder to make than plutonium. Plutonium-239 can be chemically separated from spent reactor fuel. Chemical separation could be done by solvent extraction or ion exchange. HEU production requires more work, equipment, and energy. The desire to build a nuclear bomb may be more important than the requirement for a certain amount of fissionable material. Most nations that could build nuclear bombs have chosen not to, but rather to establish alternative security arrangements. Thus, it may be more important to focus on a nation's security concerns than its technological capacity (36).

[2] Plutonium-239 does not exist in nature but is extracted from spent uranium fuel that has been irradiated in a nuclear reactor.

[3] "Disposition" means any of a number of steps from storage to disposal that may be followed after the nuclear material is removed from warheads.

U.S. DEPARTMENT OF ENERGY

Maintenance employees working in supplied protective oxygen suits tend to plutonium stored at the Rocky Flats Plant.

worldwide and in other industries that use nuclear materials but not in weapons grades. Most notably, large quantities of spent fuel from power reactors contain significant amounts of plutonium (in low concentrations). Taken as a whole, the worldwide tonnage of plutonium in commercial fuel is many times the amount in weapons grade.[4]

In countries other than the United States, some commercial plutonium is extracted routinely from spent fuel and used in commercial nuclear power plants or stored in anticipation of its use as fuel in existing reactors or new advanced reactor designs. In general, such commercial plutonium is kept as the oxide rather than the metal form used in weapons. These countries have been pursuing new generations of advanced plutonium-fueled reactors. Most of the programs in other countries, however, have experienced difficulties, and existing operating capacity is low (3).[5] In the United States, no plutonium reprocessing is done. The importance of these facts for weapons materials disposition is that the commercial needs and uses of plutonium worldwide could affect decisions about the future use of plutonium from dismantled weapons.

Nuclear warhead materials taken from dismantled U.S. weapons include, but are not limited to, plutonium pits placed in containers and stored in bunkers at the Pantex Plant, beryllium and "secondaries" returned from Pantex and housed at the Oak Ridge Y-12 Plant, and HEU also housed at Y-12. These are all considered to be in temporary or interim storage. Long-term or permanent solutions to the disposition of these materials await policy decisions by the President and Congress.

This chapter focuses on plutonium and HEU, although the disposition of many other materials from dismantled warheads is also of concern. Plutonium removed from warheads is generally given the most attention because it is a principal building block of nuclear weapons; it poses a great proliferation risk; and it represents a significant health, safety, and environmental problem. HEU poses similar problems and risks, but it is considered a simpler disposition problem because technology exists to modify and use it in many commercial nuclear reactors.

Preliminary planning efforts directed toward disposition decisions for these materials are under way within the Department of Energy (DOE), the Department of Defense (DOD), and some other agencies. Several task forces have been investigating plutonium and uranium inventory projections, and attempting to estimate what portion of these materials are to be held (stockpiled for possible future weapons) and what portion may be surplus (79). Task forces within these agencies are also investigating certain technical options for disposing of surplus materials. In addition, DOE has been preparing plans for reconfiguration of the Nuclear Weapons Complex and is in the

[4] The different isotopic content of plutonium from commercial spent fuel makes this material more difficult to convert for weapons use.

[5] Countries with advanced, plutonium-fueled reactor programs include Japan (Fugen, Joyo, and Monju reactors), France (Phenix), Britain (PFR reactor), Russia (BN-600), and Kazakhstan (BN-350). With the exception of the Monju, which is scheduled to start up soon, the continued operation of existing plutonium-fueled reactors is uncertain.

process of developing a Programmatic Environmental Impact Statement for this reconfiguration that is to include consideration of both interim and long-term storage of plutonium pits from warheads (12). Assumptions about the future mission of a reconfigured Weapons Complex, however, have not yet been publicly presented by the Federal Government.

Several DOE-sponsored studies have focused on long-range options for plutonium disposition. High-tech approaches for "burning" plutonium in advanced reactors have been given attention in recent studies,[6] as has irradiation of plutonium as mixed-oxide (MOX) fuel in reactors that are more closely related to those currently in operation. Other work covers plutonium pit storage for moderate to long-range time frames and investigations of techniques for turning plutonium into a form suitable for disposal as waste. In addition, many experts continue to debate the question of whether plutonium is a valuable asset with beneficial uses or a major liability to be disposed of in the safest and most secure way (6,16,18,26,30).

It seems clear that in the future, the nuclear weapons enterprise must pay attention to materials management and the development of long-range disposition options. Consideration of all approaches to disposition must include a rigorous examination of potential impacts on human health and the environment. Disposition scenarios should include comprehensive plans for procedures and equipment required to protect worker and community health and safety, minimize waste, manage the waste produced, and prevent the release of toxic materials. The work is complex and requires both technical excellence and management expertise. The tasks will require many decades, and the consequences will last for centuries. Capable and enduring institutions are needed to ensure success. The following sections address the options for storage and ultimate disposition of plutonium, and approaches for the disposition of highly enriched uranium.

OVERALL DISPOSITION CONCERNS

Even though an official decision has not been made, some portion of the inventory of plutonium pits that will soon be in temporary storage is likely to be deemed excess or surplus (not needed for weapons). Current studies by the Department of Energy and others on disposition options make the assumption that about 50 tons of weapons plutonium could be available in the future for other uses or for disposal.[7]

In the same manner, DOE has not officially declared that any U.S. weapons-grade HEU is surplus to the needs of military programs. However, current plans indicate that between 25 and 100 tons may become available for other uses in the future.[8]

In early 1991 the Department of Energy established a task force on plutonium strategy[9] to plan for future needs and programs to manage plutonium under DOE custody. Since then, however, world events have forced a rethinking of DOE's plutonium strategy. The task force has had

[6] The term "burning" refers to irradiation and partial or incomplete fissioning, rather than complete destruction of plutonium. Some concepts envision extensive recycling of plutonium in systems that could eventually result in near-complete destruction of most of the plutonium, but these require considerable research and testing and will generate fission products and other radioactive high-level waste.

[7] DOE has recently issued the unclassified statement: "Up to approximately 50 metric tons of plutonium will (or may) become available by about 2005 from Defense Programs (DP) inventories for use in reactors and for other civil (unclassified) purposes. Part of that material will be provided from retired weapons and part from other DP inventories" (56).

[8] DOE has recently issued the unclassified statement: ". . . under some planning scenarios substantial quantities of HEU (25 to 100 metric tons) may become available over the next 5 to 10 years and such quantities may be allocated to civil use" (56).

[9] This internal DOE group was organized under the Office of Weapons and Materials Planning within Defense Programs and reports to the Deputy Assistant Secretary for Military Applications. The data used and the materials projections made by the task force are largely classified. None of the work is subject to outside review or public scrutiny. The product of the task force is still in internal draft form and unavailable for public distribution.

to take into account actual weapons retirements and plans for future retirements.

The task force has identified and categorized plutonium in the DOE inventory; it has made inventory projections based on an expanded weapons retirement program. Based on internal interpretations of stockpile plans, the task force has projected plutonium requirements for both future weapons programs and other uses and it has identified some options for future plutonium management. The plutonium material considered by the task force includes pits from dismantled warheads, pits in the process of being reworked for the stockpile, and materials such as metals, oxides, and residues that are left over from past production operations (80).

The task force completed its initial work early in 1993. Besides the plutonium in weapons still in DOD custody, the task force identified five categories of plutonium material, which it defined in terms of intended use or disposition:

1. plutonium in active use in the weapons production program;
2. a strategic reserve of plutonium for future weapons programs;
3. a reserve of plutonium for future, nonweapons programs;
4. a national asset reserve of excess plutonium for unspecified use; and
5. plutonium residues for treatment and disposal.

Disposition Approaches

Methods for plutonium disposition present a variety of difficult technical, regulatory, economic, environmental, political, and public policy questions. There is no consensus in the United States today about what to do with plutonium from weapons, and there is some question whether one or more sites can be identified at which the public will accept long-term plutonium storage.

The ultimate disposition of plutonium from dismantled nuclear weapons represents a problem without a ready technical solution. Ideally, options would be judged in light of how well they may accomplish relevant national goals and policies arrived at after public debate. Yet, to date, technical debate among experts about the merits and limitations of alternatives for managing surplus plutonium is taking place before important decisions about national goals and policies have been made by the Federal Government. Therefore, it is difficult to measure proposed options against goals and policies in an informed public debate. In addition, there is a wide diversity of opinion about how soon a decision regarding ultimate plutonium disposition (at least in the United States) needs to be made.

The Office of Technology Assessment (OTA) has gathered certain data and analyses about the plutonium disposition issue, and has held workshops to explore various approaches and their relative merits. The advantages and disadvantages of many options are being investigated by groups within or supported by the Department of Energy and through studies by the National Academy of Sciences. Conclusions and recommendations from these and other technical study efforts will probably be reached during the next year or two. At the same time, it will be important to make progress on defining national goals and developing a process to address the national security, political, environmental, and social impacts of various technologies that could be used.

Decisions in the United States about disposition of plutonium from warheads might also consider the disposition of plutonium residues and other special nuclear materials in various forms that were manufactured for either weapons or commercial use. Also, if certain technologies are pursued, it might be useful to consider whether they could have merit and application in other countries, particularly Russia and other members of the former Soviet Union. Another factor that might be considered is the future of the nuclear power industry. Civilian plutonium that has been separated from spent nuclear fuel in other countries might also be considered when

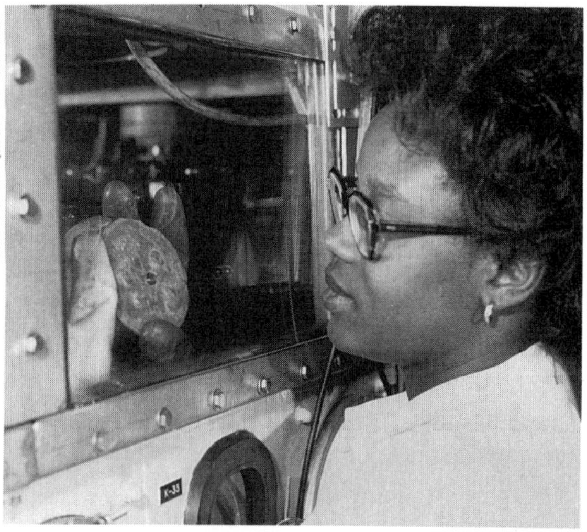
U.S. DEPARTMENT OF ENERGY

An inspector is reviewing a plutonium button at the Rocky Flats Plant. This is an example of the current practice employed for protecting workers during processing operations.

planning long-term disposition. Even though U.S. national security goals might be limited to controlling materials from other country's warheads in the short term, various commercial nuclear power activities could have a long-term impact on uses and demands for the same or similar materials.

The Russian plutonium situation should be carefully considered. For example, the issue of whether Russia will extract plutonium through reprocessing of spent fuel in the future could be influenced by U.S. decisions to pursue certain technologies for plutonium disposition.[10] Some believe that commitments by Russia and the United States to reduce nuclear arsenals have created an opportunity to reach agreements to stop the production of more plutonium worldwide as part of a general effort to limit the proliferation of nuclear weapons (3,16).

Recent studies that address the issues surrounding plutonium disposition have generally focused on one or more of the following:

- retrievable plutonium *storage*, with or without a change in form, for periods up to 100 years or more (possibly as pits, metal ingots, or oxides);
- *processing* (''burning,'' ''transmutation,'' ''annihilation'') of plutonium to destroy some portion or dilute and contaminate it, rendering it more proliferation resistant as spent fuel (this includes use as a fuel in existing or new civilian nuclear power reactors, or in special dedicated government facilities); and
- disposal of plutonium as *waste*, with or without some suitable change in form, with possible addition of high-level waste or specific fission products (e.g., cesium-137).

Each category has variations with unique implications, and the categories are not mutually exclusive. Most will be necessary to a greater or lesser degree at some time in the future. Some storage is required for all categories, but the time frame could vary significantly among them. Minimal processing is probably also necessary if only to maintain stability for long-term storage. The extent of processing could also vary greatly. In the end, some long-term disposal will be needed either for unconverted materials or for residuals and waste. Table 4-1 summarizes the categories covered in this chapter. Figure 4-1 illustrates the various paths that could be followed after dismantlement to dispose of plutonium from warheads.

It is virtually impossible to judge or compare most plutonium disposition technologies as reported in the literature unless one can be sure they are being evaluated using the same original assumptions. Some investigations of plutonium disposition options begin by assuming that the material no longer used for weapons is still a ''national asset'' and that research should be directed at extracting the greatest benefit from

[10] Although the United States has publicly announced that it stopped plutonium production in 1988, some U.S. investigators and Russian officials state that Russia continues to operate plutonium production facilities (16, 47).

Table 4-1—Summary of Selected Plutonium Disposition Approaches

Category	Approaches discussed in OTA report	Comments
Storage	Existing storage of pits. New long-term storage facility.	Some storage will always be necessary.
Processing	Mixed-oxide fuel reactors. Advanced metal reactor. High-temperature gas-cooled reactors. Accelerator-based converter.	All processing options require development, and their feasibility and applicability depend on the results of such development.
Waste disposal[a]	Deep geologic disposal in containers after vitrification to form glass logs. Sub-seabed disposal. Disposal in space. Underground detonation.	Waste options require some technical development and may be difficult to support without convincing economic arguments.

[a] Waste disposal will eventually be necessary even if a processing option is chosen because no processing method can totally destroy all residuals of plutonium contamination from waste streams.

SOURCE: Office of Technology Assessment, 1993.

this asset. Such benefits could be either to produce energy or to provide the impetus for a future large-scale nuclear power economy. Other investigations make the assumption that plutonium is a liability and that systems should be sought that would most effectively destroy it or render it unusable.

Regardless of which long-term approach is pursued, storage for some period of time will be required for plutonium from dismantled nuclear warheads. Retrievable, monitored, and secure storage is inevitable while warheads are being disassembled and other long-term options such as processing or disposal as waste are being investigated. The period could last from one to many decades. There is sufficient existing technical knowledge about plutonium storage to have reasonable confidence in performance (technical, economic, safety, and environmental).

The conversion of plutonium to mixed-oxide (plutonium and uranium) fuel for use in light-water reactors is also considered by most experts to be technically feasible[11] in the near term. The basic technology to develop a facility for vitrification of plutonium, perhaps mixed with other radioactive products, to form a waste is also available.

Other technologies for plutonium disposition require various amounts of research and development. Some preliminary investigations are under way, but resources are limited. It will be very important to follow even these preliminary investigations, however, and understand their conclusions. Such conclusions will always involve compromise (among factors such as cost, time, and uncertainty); thus public debate about national benefits and costs will be important to their acceptability. Any decision about disposition must inevitably take into account the length of time that storage would be acceptable from both technical and security points of view, so that adequate research, development, and testing of other technologies (including environmental impact analyses) can be carried out.

To evaluate plutonium disposition options and select the most appropriate, it will be necessary to establish clear and measurable criteria. The criteria must specify objectives to be achieved and some means of measuring how well they are achieved. The criteria should be given relative

[11] This technology has been demonstrated and used in other countries.

Figure 4-1—Warhead Dismantlement and Plutonium Disposition Scenarios

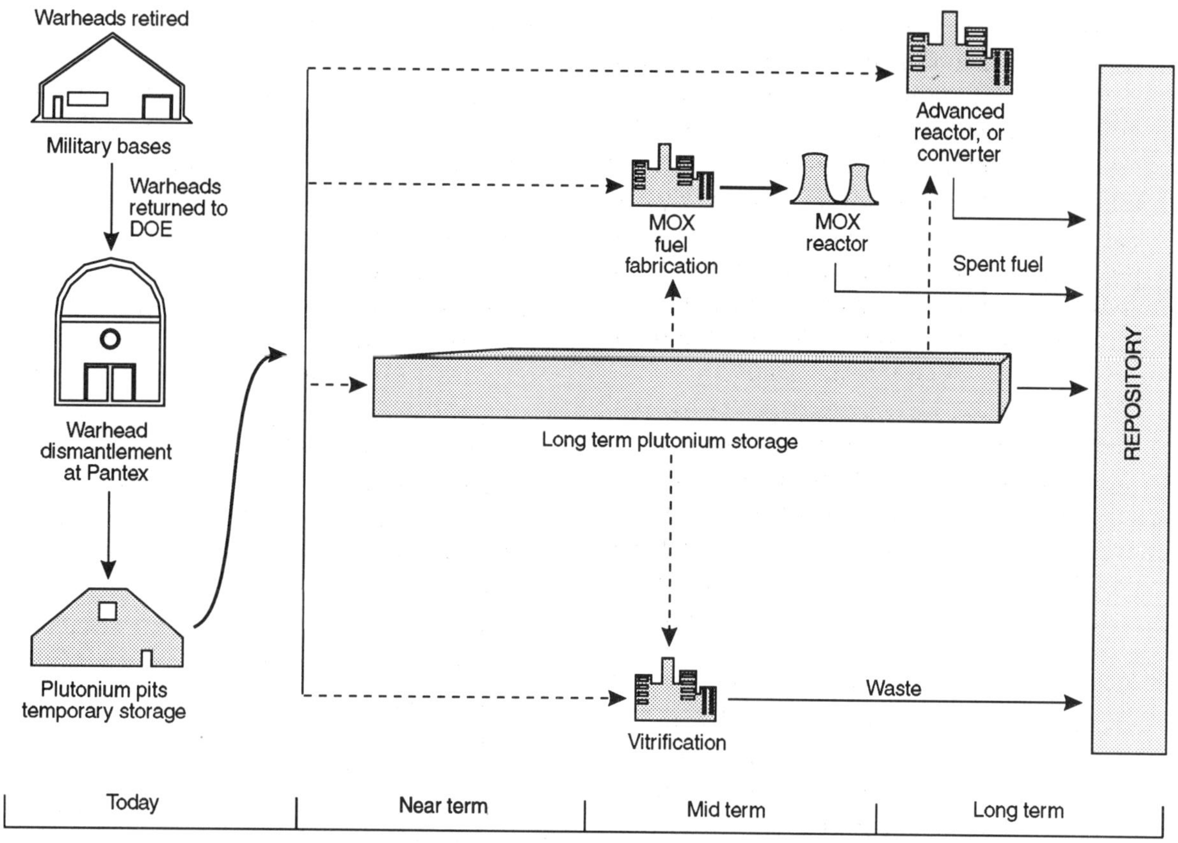

A dotted line indicates alternative disposition paths.

SOURCE: Office of Technology Assessment, 1993.

weights or listed in priority order. Establishing such criteria will be difficult, but important. To reflect a public consensus about national goals, they must also involve the public in the decision-making process.

Criteria for Judging Disposition Approaches

Individual researchers have developed their own notions about what criteria should be considered and which should be most important. A number of such criteria can be found in studies (3,4,20,23,48) whose principal results are discussed later in this chapter. OTA's analysis indicates that the criteria listed below are among the most important. This list is not necessarily complete, but it is a starting point. The items are not necessarily listed in order of priority. These criteria are based primarily on the oft-stated assumption that world peace and security will be enhanced if nations of the world reduce their nuclear weapons stockpile; prevent the materials from being released into the environment; render such materials as harmless as possible for future generations; and prevent proliferation of materials that might be reused for new weapons.

1. Security (including verifiability and proliferation resistance). Each approach must be judged on how well the material is controlled and protected from theft or other diversion. It

is necessary both to protect the material from possible terrorist actions and to prevent certain nations from receiving such material through either overt or covert means (69). If a future international agreement on storage, use, or disposal of plutonium is sought, acceptable means of verifying compliance will have to be established. Thus, an approach must also be judged on how well the amounts and forms of plutonium can be accounted for, measured, and controlled. An option could also be judged on how quickly the material could be converted to a more proliferation-resistant form.

2. Near-term health and safety risks. As discussed throughout this report, each approach must be judged on how well the health and safety of workers and the public are protected throughout the time the material is stored, moved, handled, and processed. Risks of human exposure to plutonium and other toxic materials are of primary importance. Risks of accidents, as well as exposures that may be associated with routine operations, must be considered. It is also important to consider in great detail the many complex steps usually involved in certain plutonium processing options.
3. Environmental and long-term health risks. All approaches must be measured by the degree to which environmental protection can be ensured and future exposures of humans to toxic materials can be prevented over long time frames. Because these materials have very long half-lives, the viability of a geological repository for long-term disposal is a prime consideration if plutonium is to be disposed of as waste.
4. Technical availability and feasibility. Most available work on disposition has included preliminary evaluations of technical feasibility. When options are compared, however, it will be necessary to realistically assess the status of development of some very complex systems; the nature of technical uncertainties associated with each; the possibility of technical failures; and the time needed to justify, fund, design, build, test, license, and operate a full system.
5. Economics and cost. All options will be expensive, but to compare them it will be necessary to treat all costs on an equivalent and consistent basis. The options must be measured by a comprehensive evaluation of relative costs including the degree of uncertainty associated with each cost estimate. Potential benefits such as the value of electricity produced should be a factor, as should potential costs from accidents or environmental releases. Cost recovery should be measured in a consistent and comprehensive way. Researchers have presented some cost data in various studies to date, but none are of sufficient quality that comparisons among options would be fruitful.
6. Political and public acceptance. The consideration and debate of each option must include adequate involvement of the general public, experts, and various political interests. To satisfy this criterion, it will be necessary to consider public concerns—to understand how an option can be presented to the public, how public opinion will be formed, how public input can be incorporated into decisions, and how the public will measure benefits and costs.
7. International political impacts. Any choices made by the United States regarding the storage and disposition of plutonium and HEU recovered from dismantled nuclear warheads will have an impact on the way other nations approach this issue. Considerations could include the following: Will any option selected assist in ensuring that Russia will permanently destroy surplus weapons plutonium? Will it assist in securing a commitment from Russia to prevent the further separation of plutonium from reactor discharge materials? Will it assist and reinforce the U.S. position to discourage the separation and recycling of commercial plutonium worldwide, and to find an ultimate solution to the disposition of commercial plutonium. Finally, should any international reciprocity be considered for plutonium and HEU disposition options?

Connections Between Civilian and Military Plutonium

Various analyses place different emphases on individual criteria. Almost all studies to date, however, regard proliferation resistance as a critical factor, and therefore the economic benefits or costs of different plutonium disposition options may not be overriding factors in the selection or elimination of any option (5).

The continuing production of new plutonium is also a factor used by some in evaluating schemes for disposing of existing weapons plutonium. Reports indicate that plutonium production and separation (reprocessing) continue in Russia. The rationale for running these reactors is that they are required to produce energy for the associated towns (67). Russian officials claim that reprocessing is continuing because it would be unsafe to store spent fuel from certain reactors or because it is part of a continuing effort to develop advanced plutonium-fueled reactors (16). Commercial plutonium reprocessing is also expanding (or planned to expand) in other countries (3). Although the United States has adopted a policy of not reprocessing any commercial fuel to recover plutonium, some other countries have pursued a nuclear policy that calls for reprocessing spent nuclear fuel to separate and recycle plutonium in reactors.

Worldwide, the civilian nuclear industry has already separated more than 100 tons of plutonium from spent fuel (67). Some of this has been "recycled" in various types of reactors, but the remainder is in storage. Most of the civilian international industry for plutonium separation is in Britain and France, but Russia has facilities and Japan is constructing some. These countries plan to separate another 200 tons over the next decade (67). The additional 200 tons to be separated is covered by contracts with reprocessing plants in Great Britain and France (3). Plans also call for this plutonium to be returned to the originating countries and thus entail a significant expansion in the handling, transportation, and circulation of plutonium, which will add to global proliferation, safety, and environmental risks (3).

Large amounts of separated civilian plutonium could be a factor in decisions about technologies that might be developed to convert plutonium from warheads. Since substantially more plutonium is available in spent fuel from civilian power reactors than is likely to become available from warhead dismantlement, some argue that it would be logical to consider the problem of weapons and civilian plutonium together, rather than separately (45). Others argue that the storage and production of separated civilian plutonium should be controlled in a manner similar to military material (16).

The control and management of plutonium from both weapons and civilian power reactors could be based on the same nonproliferation concerns (3). Some researchers believe that initially both must be stored under international safeguards and that there should be a verified ban on separation of any new commercial plutonium. Researchers also argue that the principal reason for current reprocessing and recycling activities in Western Europe and Japan is institutional inertia rather than economic benefit, and that this increased plutonium activity is unjustifiable on security, economic, or environmental grounds (3).

On the other hand, although recognizing that the large amount of civilian plutonium represents a serious proliferation problem, some think that there are both political and technical reasons for proceeding expeditiously with a permanent solution to the disposition of surplus military plutonium even if a solution to the civilian plutonium issue is not currently available. Weapons-grade plutonium comes in the best form for warhead construction. It is also in a form that can be modified more readily for certain disposition options such as conversion to oxide and glassification with high-level waste. Finally, timely actions by the United States to permanently dispose of surplus weapons plutonium may strengthen its ability to influence Russian disposition actions, and will emphasize the U.S. position regarding the disposition of commercial plutonium and its world leadership role in nonproliferation (24,30).

PLUTONIUM STORAGE

OTA's analysis indicates that storage of most of the plutonium from weapons for a few decades at least is the most likely outcome of the plans and programs now under way in the United States. Other options for disposition will require considerable research, development, and testing before they can be implemented, and they must surmount significant technical and political hurdles to meet other criteria. In addition, the Federal agencies involved in making disposition decisions are generally reluctant to dispose of plutonium permanently because of the enormous cost and effort expended to create this material.

It is important to treat storage with great care and concern. Safe, secure storage requires attention to design requirements and to all factors that can affect protection of human health and the environment. It should be remembered that past inadequate practices in managing radioactive waste from weapons production have led to the vast environmental problems now existing in the Nuclear Weapons Complex (65). No one wishes to repeat those mistakes, but avoiding them will require that difficult decisions be made about providing adequate storage facilities and the best protection possible under future storage conditions.

It is important to begin soon to prepare plans for mid- to long-term storage of plutonium from dismantled weapons.[12] DOE is exploring storage options through its work on reconfiguration of the Weapons Complex and its plutonium task forces, but these efforts are not well coordinated. Among the factors to be considered initially are the size of a facility (number of pits or other forms to be stored); whether other plutonium forms and residues should be accommodated as well (there are now substantial quantities of plutonium in various forms throughout the Weapons Complex); the estimated life of a storage facility; and any additional capability required, such as the ability to handle and maintain some pits or classes of pits that need attention over time.

Current Efforts

DOE has the responsibility to evaluate all relevant issues pertinent to plutonium storage. Plutonium pits from dismantled warheads are currently considered by DOE to be in interim storage (6 to 10 years)[13] at Pantex. Because the capacity of the Pantex bunkers is restricted, DOE has prepared analyses of the safety and feasibility of expanding that capacity to a maximum of 20,000 pits. An Environmental Assessment (EA) has been prepared that incorporates the results of these analyses. In the EA, DOE discusses the potential of other Nuclear Weapons Complex sites as interim storage facilities (see table 4-2). Some of these alternatives may be considered in connection with siting a long-term plutonium storage facility in a reconfigured Nuclear Weapons Complex.

The conclusion from DOE's initial efforts is that storage of plutonium pits at Rocky Flats or Hanford is neither reasonable nor cost-effective because current plans call for environmental restoration and no further use of these sites for any production purpose. Another alternative evaluated is to move the pits to one of the Weapons Complex sites not planned for closing, such as Savannah River. However, efforts to expand the

[12] Any major new Federal facility to be built will require a long time (more than a decade with current DOE procedures) from initial plans and concepts to actual completion and first use (43). Some experts claim that an adequate storage-only facility similar to Pantex could be built in a much shorter time and at a cost of less than $100 million (58), but none of these estimates has been well developed or documented.

[13] Many believe that DOE will not be able to provide a site and facility to replace the Pantex bunkers within 6 to 10 years. See appendix A for a discussion of the current proposal to expand plutonium pit storage capacity at Pantex and the reaction to this proposal by the local community, the State of Texas, and other citizen groups.

Table 4-2—Alternatives Considered by DOE for Interim Plutonium Pit Storage

Possible storage site	Storage capacity available for plutonium pits	Storage capacity available for other plutonium forms	Issues relevant to this activity
Pantex Plant (Texas) Single-layer configuration	Standard single-layer configuration is capable of providing interim storage for only 6,800.	Storage of other plutonium forms has not occurred and is not planned at the present time.	The concrete storage bunkers were built during World War II to protect conventional weapons and munitions from bomb blasts. Although some pits are already stored in these igloos, DOE is evaluating the potential impacts of extending this storage.
Multiple pit stacking	If approved, the proposed multiple pit stacking configuration will provide interim storage for up to 20,000 plutonium pits from disassembled weapons.	None	Environmental and safety documentation is being prepared and reviewed.
Hanford Site (Washington)	Facilities at this site are capable of providing storage capacity for more than 10,000 plutonium pits.	Existing facilities might enable the storage of nearly 20 tons of plutonium in forms other than pits.	Requests for funds to upgrade facilities for plutonium storage would appear to be in conflict with DOE's change in policy from defense missions to those of environmental restoration and waste management.
Los Alamos National Laboratory (New Mexico)	Existing facilities have limited capacity for storing plutonium pits. One facility currently under construction could provide storage for about 200 plutonium pits; however, larger capacities could be developed.	Storage for other plutonium forms is available but largely limited to certain forms such as plutonium oxide.	Cost of modifying facilities under construction or upgrading existing facilities is high.

Table 4-2—Alternatives Considered by DOE for Interim Plutonium Pit Storage (Cont.)

Possible storage site	Storage capacity available for plutonium pits	Storage capacity available for other plutonium forms	Issues relevant to this activity
Rocky Flats Plant (Colorado)	The capacity currently available for near- and long-term secure storage of plutonium pits is limited.	The space available for providing environmentally safe and secure storage is sufficient merely to accommodate the plutonium scrap, residues, and waste generated by the plant's past plutonium processing and current cleanup activities.	Storage of additional pits or other plutonium forms is a remote possibility because of the extensive costs and difficulty associated with facility and equipment upgrades. Addressing relevant environmental and safety problems would also be difficult.
Savannah River Site (South Carolina)	Use of existing facilities could provide storage space for up to 1,100 plutonium pits from nuclear weapons disassembled at Pantex.	Storing plutonium in forms other than pits is possible because current activities involve the storage of plutonium oxide and plutonium-rich residues originating at the site.	Without modifications to some facilities, storage capacity for plutonium pits may be further reduced by future shipment of plutonium materials and residues from other DOE sites.
Military bases	Although viewed by DOE as facilities with little potential for near-term plutonium pit storage, the possibility of using certain bases for long-term storage seems promising to DOE. The estimated capacity for pit storage has not been determined.	The possibility of using military bases to store other forms of plutonium has not been suggested or evaluated to date.	Many experts believe that most military facilities were designed for weapons storage only and are unsuitable for plutonium pits. Factors to be evaluated include institutional arrangements and costs associated with inspection, security, and surveillance requirements.

SOURCE: U.S. Department of Energy.

storage of plutonium pits at Pantex will also have to be continued since, according to DOE, alternatives such as the Savannah River Site would not independently provide the necessary capacity soon enough (73).

DOE also considered certain military bases as potential candidates for interim storage of plutonium pits. No detailed evaluation of converting facilities from weapons storage to pit storage was done, and the military services indicate that they do not have excess capacity at any of the candidate bases (73). Higher costs and additional logistical considerations could make storage at military bases difficult to implement[14] (34).

DOE is currently evaluating approaches that could lead to replacement of the current oversized Nuclear Weapons Complex by a new one in the year 2000 and beyond; this future complex is commonly referred to as Complex 21. The Los Alamos National Laboratory is preparing plutonium storage design guidance to be used in the design of a plutonium storage facility for Complex 21. In addition, private contractors are providing technical support for a conceptual design and cost estimate for each alternative under consideration (72). DOE is using three major assumptions in regard to a plutonium storage facility:

1. It must be a modular design with remote handling capability to reduce worker radiation exposure;
2. It must consist of storage vaults and welded containment vessels that minimize risk of intrusion; and
3. It must provide adequate capacity for safe, secure, long-term storage for projected amounts of plutonium pits, metals, oxides, and other stable forms (42).

U.S. DEPARTMENT OF ENERGY

Plutonium pits from dismantled warheads are temporarily stored at the Pantex Plant in bunkers like the ones shown here.

Other characteristics expected in the final design are that it must be self-contained, although it could share other support facilities located at the site, and it would be constructed at grade level rather than underground. The central advantage of adopting a modular approach is that modules can be added as required, thus eliminating potential capacity limitations for the plutonium form in question (12).

DOE is evaluating the storage of plutonium and highly enriched uranium at separate sites, along with the possibility of a single facility capable of storing both. The results of engineering and cost evaluations are expected to be published in early FY 1994 (9). The effect that the size of weapons stockpiles may have on future plutonium storage needs is also part of this continuing evaluation.

The storage facility design concept under consideration by DOE includes a Class I vault

[14] Several researchers have also pointed to the Manzano Mountain facility at Kirkland Air Force Base as a possible alternative for plutonium pit storage. Pits were customarily stored at this facility, especially during the 1940s and 1950s when Manzano Mountain was considered the primary assembly area for nuclear weapons in the United States. Weapons assembly and plutonium pit storage are no longer conducted here. Before Manzano Mountain could be considered adequate for storing plutonium pits, however, several important issues (and their cost and time implications) need to be addressed: 1) the analyses and design modifications required to meet modern environmental and safety standards for plutonium pits; 2) the capacity available for pits; 3) programs to protect the health of workers; 4) programs to monitor and control radiation; and 5) maintenance associated with storage and security operations (50).

storage system to meet upgraded security and safety standards that are not found in current facilities such as the Pantex bunkers. One example of existing Class I facilities in the Nuclear Weapons Complex is the storage vaults recently built at the Savannah River Site.

Modifications to an early proposed design indicate that the final structural design for a long-term plutonium storage facility is still evolving. For instance, early designs assumed a 50-year life to address DOE's plutonium storage problem. This was found inadequate by some DOE reviewers, and a new structural design for a 100-year facility is now being proposed—with plans for replacing computer hardware and other special equipment every 25 years (34). Figure 4-2 illustrates the current design features being considered for this storage concept.

The recent reduction in the nuclear weapons stockpile, the closing of key processing facilities, and the downsizing of the Weapons Complex have also significantly changed DOE's approach to long-term plutonium storage. DOE is preparing a Programatic Environmental Impact Statement (PEIS) for the Weapons Complex reconfiguration. One of the objectives of this study is to evaluate engineering and environmental approaches to replace the current Weapons Complex with one that is simpler, more environmentally safe, and less expensive to operate. The PEIS will also evaluate strategies for long-term plutonium storage.

As part of its efforts to reconfigure the Nuclear Weapons Complex, DOE created a Complex Reconfiguration Committee in 1991, with senior representatives from DOE, DOD, and the National Security Council. The committee considered such aspects as future stockpile needs, long-term production and maintenance requirements, environmental needs, and options for existing or new facilities (71). DOE had planned to issue the PEIS in August 1993, but changes in the weapons stockpile resulting from recently signed arms reductions agreements have led to revisions of their schedule (4,9,46). In July 1993 DOE issued a revised notice of intent to prepare a PEIS that explained the new conditions that caused this revision and provided a new list of options to be considered.[15] DOE stated that since February 1991 when DOE originally announced its intent to prepare a PEIS for reconfiguring the

Figure 4-2—Conceptual Design of a DOE Plutonium Storage Facility

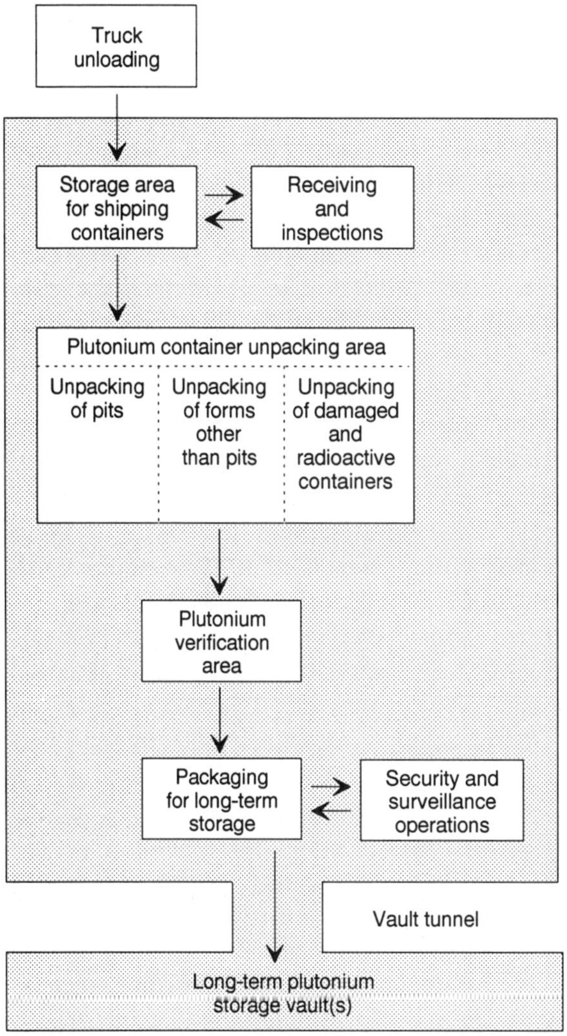

SOURCE: U.S. Department of Energy.

[15] *Federal Register*, vol. 58, No. 140, July 23, 1993, pp. 39528-39535.

Weapons Complex, conditions have changed in such a way that impact the requirements for the new complex. DOE's proposed changes in the PEIS reflect that the future nuclear weapons complex can be even smaller than originally envisioned, and also reflect the increased importance associated with stewardship of existing nuclear materials.

One major change in scope is that DOE considers it unreasonable to have plutonium component fabrication at a different site than storage facilities. Therefore, DOE proposes that all alternatives under consideration will have storage, processing, analysis, and fabrication operations co-located. For the function of nuclear materials storage, processing, and component fabrication, DOE now proposes three alternatives—constructing new facilities, modifying existing facilities, and no action. If new facilities are constructed, five alternative sites will be evaluated—Idaho National Engineering Laboratory, Savannah River Site, Oak Ridge Reservation, Pantex, and the Nevada Test Site.

Public scoping meetings on the revised PEIS will begin in the fall of 1993, and a plan and schedule will be announced later. If a decision is reached to build a long-term plutonium storage facility, some estimate that constructing such a facility will require at least 10 years (42). It now takes DOE more than a decade to obtain funds and build new budgeted projects, even if the technology is tested and proven (43).

Designing for plutonium storage beyond a few years is a relatively new concept within the Weapons Complex. For many years, plutonium pits recovered from warheads were stored only briefly at Pantex and then shipped to Rocky Flats where they were processed for recycling into new weapons. Scrap plutonium metal and other residues were stored at generating sites with the intention of recovering the plutonium when production ceased or when more effective recovery technologies become available. The current very costly and complex challenge to dispose of plutonium residues from past operations at the Rocky Flats Plant (see box 4-A) illustrates how past practices without attention to environmental protection have created massive waste management problems with no adequate, feasible, or practical solution. Future planners and designers should heed this lesson carefully.

Design Considerations for a Plutonium Storage Facility

OTA's analysis indicates that certain considerations will affect the design parameters for a plutonium storage facility. Box 4-B lists the types of technical and related analyses that would be required as part of any facility design. Additional considerations will also be important in designing a plutonium storage facility.

For example, it will be important to identify a time period within which pits can be stored safely without further processing. If intact pits are to "sit on the shelf" for a defined period of time, the pit casing and sealed storage drum could obviate the need for immediate processing. However, once the design life of the container or casing is reached, adequate processing capability will have to be provided (51). There is also a need to assess the chemical and physical stability of the plutonium materials to be stored (e.g., pits, metals, oxides, glass, ceramics), and to define the sizes and concentrations of materials selected for storage so as to determine the space required for containment and criticality control. Although it may be appropriate to store plutonium as pits for a defined temporary period, further study is required to determine any limiting factors for long-term pit storage.

Another design consideration is the need to evaluate opportunities for the use of remote handling technologies (e.g., robotics) in storage and maintenance areas (51). The selected containment system (e.g., drums, vessels, vaults) should be designed in a way that facilitates inventorying stored materials with minimum radiation exposure of workers. There is also a need to protect workers against plutonium particle exposure.

Box 4-A—Past Experience with Plutonium Processing at the Rocky Flats Plant

For nearly 40 years, the fabrication of weapons parts from plutonium metal took place at the Rocky Flats Plant (RFP) in Colorado. The limited efficiency and complex nature of the processes employed resulted in the generation of a large inventory of plutonium-contaminated scrap materials and residues ranging from processed plutonium materials that failed to meet weapons design specification, to scraps from shaping and machining, to contaminated items used during processing, to residues and waste. The high costs of plutonium production and the potential for future economic recoverability have, in the past, motivated DOE to store both scrap materials with high plutonium content and residues having lower plutonium levels. To date, plutonium can be found in more than 100 different types of residue at Rocky Flats. Other DOE sites where plutonium scrap and residues are found are Los Alamos National Laboratory, the Savannah River Site, and Hanford.

The handling of plutonium residues at Rocky Flats has traditionally been under the authority of DOE's Office of Defense Programs (DP). On January 15, 1993, however, a Memorandum of Agreement was signed between DP and the Environmental Restoration and Waste Management (EM) Program to share the responsibility for managing plutonium residues.

Under this agreement, DP continues to be responsible for managing RFP's plutonium-rich residues. Materials in this category include plutonium weapons parts, metal buttons, scrub alloy, and plutonium oxides. Limited processing is usually needed for these residues to meet transportation and long-term storage requirements. The plutonium present in these materials may be recoverable with existing processes or with technologies expected to be developed in the future.

With the signing of the January 15 agreement, EM is now responsible for the processing, storage, and safe disposition of more than 90 percent of the volume of plutonium residues at Rocky Flats. The average plutonium content is 2.6 percent by weight. In total, EM will manage nearly 7,300 containers of varying shapes and volumes containing from 6,000 to 7,000 pounds of plutonium. The presence of regulated hazardous constituents in the majority of these materials requires that their treatment and disposal be conducted according to the Resource Conservation and Recovery Act. The chemical instability associated with the majority of these materials makes their shipment to off-site storage or disposal sites without treatment or processing highly unlikely.

As part of a Programmatic Environmental Impact Statement for the reconfiguration of the Weapons Complex, Defense Programs is developing a long-term strategy of future needs, uses, and possible storage locations for plutonium. It is also reviewing several facility and technology designs and costs to provide storage for stable plutonium residues. EM is addressing problems associated with the low-concentration plutonium residues and evaluating technologies for stabilization, treatment, and storage.

SOURCES: General Accounting Office and U.S. Department of Energy.

During storage, plutonium metal (as found in pits) may oxidize and form particles small enough to be respired by humans. Even though the risk to workers of plutonium exposure during storage is low, accidents that could disperse fine particles are always a concern. In addition, if plutonium is processed (e.g., converted to oxide) or if pits are converted to small pieces, there is a risk of dispersion in forms susceptible to inhalation or ingestion. Plutonium, which emits alpha radiation, is dangerous when inhaled or ingested (see chapter 3). Also, over time, weapons-grade plutonium will form americium-241, which emits penetrating gamma radiation.[16] Since all military plutonium contains various amounts of americium, it must be handled with appropriate shielding precautions (18).

[16] Weapons-grade plutonium contains mostly plutonium-239 and smaller amounts of plutonium-241, which naturally decays over time to americium-241 whose half-life is 13.2 years.

Box 4-B—Types of Analyses Required in Designing a New Long-Term Plutonium Storage Facility

- Safety Analysis Reports that address:
  General description of principal design criteria,
  Nominal capacity considered for the facility,
  Type, form, quantities, and origins of the plutonium materials,
  Waste products generated during operations, and
  Materials handling and storage procedures, including control of decay heat, criticality safety, contamination control, and criteria for handling damaged containers.
- General operating procedures for packaging, storage, and transportation.
- Design criteria for ventilation, filtration, and off-gas systems.
- Criteria for protection of equipment and selection of instrumentation.
- Radiation protection and control measures.
- Fire and explosion protection systems.
- Requirements for containers, container repair, and maintenance.
- Procedures to be used for monitoring.
- Classification of structures, components, and systems.
- Criticality prevention and criticality factor analyses.
- Maximum radiation dose rates emitted by containment systems.
- Procedures for decontamination of personnel and equipment.
- Accident potential for normal and abnormal operations.
- Design criteria and general operating procedures relevant to security, verification inspection, and monitoring.
- Organizational structure, including functions, responsibilities, and authorities.

SOURCE: Office of Technology Assessment, 1993.

There are also broader policy issues to be considered. These include the need to evaluate security factors associated with storing plutonium at a consolidated facility as opposed to two or more locations. Preliminary analyses appear to suggest that placing plutonium in a centralized location may be more cost-effective. Each location will require significant security measures, including redundant barriers to slow down individuals who attempt to take possession of the stored materials (41). There are some advantages, however, to building two facilities, such as making international or bilateral verification easier.

It is important to consider whether a U.S. plutonium storage facility might become subject to international safeguards for verification sometime in the future. Some experts maintain that in order to minimize security, accountability, and proliferation problems, plutonium storage would best be carried out in collaboration with other nations that possess nuclear weapons (33). National security considerations also raise the question of whether verification by foreign governments or by any international organization, such as the International Atomic Energy Agency (IAEA), could be allowed in the future as the result of amendments to arms reduction treaties. Pending the development of an international plutonium and radioactive waste disposal strategy, some have suggested that the best interim solution is monitored, secure storage of surplus plutonium under bilateral safeguards (25,53). Although weapons plutonium could initially be placed under safeguards through bilateral agreements, some believe that in the long term, an international control entity such as the IAEA might better reflect a global interest in keeping these materials from weapons use (3).

In any event, some experts suggest that the facility design should enable verification inspections (46) and should accommodate possible modification of plutonium materials to meet verification requirements. These considerations would have an effect on the design, optimum number, and location of storage facilities.

Optional Form of Plutonium Storage

The ideal form in which to store plutonium depends on the goals set for storage. Different goals—such as greatest accessibility for possible weapons use in the future, highest proliferation resistance, or minimal impact on the environment and workers—may dictate different storage forms. Stability is also an issue. Some argue that plutonium metal is less desirable for storage than the more stable oxide form because fine metal pieces can ignite spontaneously if exposed to air. In addition, some claim that storage as plutonium oxide has proliferation resistance advantages compared with storage as metal (3), whereas others say that such advantages are minor (19).[17] However, the technology needed to convert plutonium oxide into its metallic form is easily accessible (39). Another point is that oxide powder may pose a greater health risk because it is more respirable.

If consideration is given to international verification and inspection of a storage facility, it would be necessary to protect weapons design information from disclosure. In this case, some changes to the pits that would modify their shape or convert them to small pieces may be desirable prior to storage. This process is commonly known as ''sanitizing'' the component. Another approach, to minimize the risk of disclosure of sensitive design information, would be application of verification measures only to sealed containers holding the sensitive materials, etc. (55). Passive nondestructive neutron and gamma-ray spectral assay procedures are sufficient for the verification of plutonium, and a combination of active neutron interrogation methods and passive gamma-ray spectral analysis could be used for HEU (55).

DOE has stated that the new Special Recovery Facility at the Savannah River Site is an existing facility with the potential to process plutonium pits into plutonium oxide. Originally constructed to transform high-grade plutonium oxide into metal buttons for use at Rocky Flats in making plutonium pits for nuclear warheads, the new Special Recovery Facility was never operated. Savannah River officials consider that reversing the intended function of the unused plant—processing pits into oxide rather than vice versa—may involve only minor design modifications. One additional function this facility could serve is to remove americium and other hazardous radioactive decay products from stored plutonium materials (34). DOE claims that the processing of plutonium pits into plutonium buttons is currently possible at the facility. One problem with the facility, however, is that it was not built to meet current environmental and safety standards, and if completely shut down, it would be very difficult to reopen under modern requirements (34).

On the other hand, storing plutonium pits in their original form may have some advantages in terms of ease of verification because each pit already represents a discrete unit and has a serial number (28). The cladding of plutonium pits was designed to have a 20-year lifetime but could probably last much longer (28).

---

[17] Some researchers have suggested that pits could be made unusable in warheads by simple means such as crushing or filling them with boron and epoxy. These approaches might deter a terrorist group but not a nation with weapons manufacturing capability, and are suggested mainly for nations other than the United States in which good security technology may not be in place.

Plutonium Pit Maintenance During Storage

If plutonium is to be stored in pit form, it will be necessary to have the capability to inspect, modify, repair, or otherwise process any pits or materials that exhibit problems. Los Alamos National Laboratory and Lawrence Livermore National Laboratory already have some capability for pit handling and processing, which is used in connection with current Pantex operations. DOE has stated that it would locate a processing facility at the same site at which a new plutonium storage facility is built (12). Options based on storing plutonium in a form other than pits (such as plutonium oxide) would also require a plutonium processing capability. Either current facilities for processing could be upgraded or new areas could be developed. It should be noted, however, that most of the facilities processing plutonium in the past were built in the 1940s and early 1950s. They are obsolete and potentially dangerous, and have been closed because of safety and environmental concerns. Lessons learned from these past operations will be valuable to developers of any new facility. In addition, these facilities often used processes and technologies that were inefficient and costly, and created large amounts of waste (40).

The Kirtland Underground Munitions Complex

There are very few good examples of high-security weapons storage facilities built recently in the United States that might be used as an example of how a plutonium storage facility might be constructed. One such facility—the Kirtland Underground Munitions Storage Complex (KUMSC) at Kirtland Air Force Base in Albuquerque—was constructed in the late 1980s according to modern standards of safety and security at an Air Force munitions complex. Although not necessarily the ideal facility for storing weapons-grade plutonium, it illustrates how modern design standards and principles

Figure 4-3—Major Design Features of Underground Munition Storage Facility at Kirtland Air Force Base

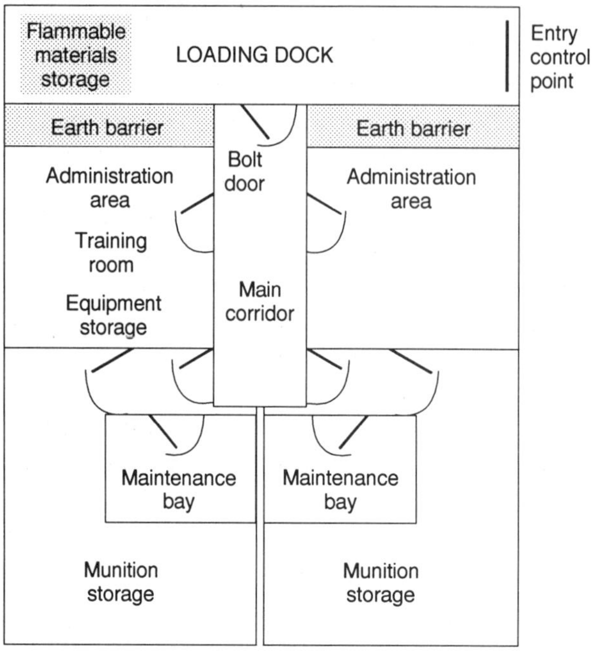

SOURCE: U.S. Department of Defense.

might be applied to a future storage facility. KUMSC consists of an Underground Munitions Storage Facility, a Squadron Operations Building, and a Utility Building covering an area of approximately 7 acres (see figure 4-3). The Underground Munitions Storage Facility comprises eight areas specially designed to sustain accidental detonation of certain high explosives and to contain detonation products. In the event of an explosion, the particular area affected is automatically isolated by the closing of blast doors; after the explosion, pressurized gases are filtered out of the explosion area and the filtered air is released to the environment. Each storage vault at the facility contains multiple storage cells with approximate dimensions of 25 feet by 100 feet. Individual cells are bounded by doors and concrete walls able to withstand accidental explosions.

Several design features have been incorporated into the Underground Munitions Storage Facility to reduce accident and security risks. Examples include: 1) limiting the use of combustible materials during construction and operation; 2) confining the number of blast doors that can remain open at any given time to one, thus exposing only two containment areas to the risk of explosion; 3) providing fire protection systems and equipment; 4) demarcating boundary lines on floor areas near walls and doors to limit the quantities of munitions that can be stored; and 5) providing only one personnel entrance/exit to the facility (gravel-filled escape tunnels secured with heavy steel plates are provided to exit the facility if the main entrance is blocked) (70). Extensive security systems protect against unwanted entry and other threats to the integrity and control of the facility.

PLUTONIUM PROCESSING

A number of studies over the past few years have looked at various processing techniques that might be applied for disposition of significant quantities of plutonium from retired and dismantled nuclear warheads (see box 4-C). In practice, it is impossible to convert surplus weapons plutonium into a substance that is essentially nonradioactive or harmless to human health and the environment. It is also difficult to transform plutonium into a material that cannot be reformed into weapons material at a later date. No existing process is available that can completely eliminate surplus plutonium, and developing new processes will require substantial research efforts and resources. However, some technologies are available in the near term to create forms that would be less usable for weapons or to eliminate some portion of the plutonium.

The language used in discussions of plutonium processing options can be difficult to interpret. Some use the term "plutonium burning" to describe the use of plutonium as reactor fuel so that plutonium levels in spent fuel are reduced over time. The same options are sometimes called "transmutation" or "actinide burning" to reflect the fact that a significant portion of the plutonium (or various transuranic species) is changed by nuclear reaction into other, shorter-lived isotopes. In more recent studies on the use of accelerators to destroy actinides, the term plutonium "annihilation" is used to depict approaches that reduce the plutonium to negligible amounts after the process is completed. Many proposals address plutonium disposition through processing. Although several current ideas have merit, it is too early in the development of most of them to compare their specific advantages and disadvantages accurately. In addition, many of the new approaches to the disposal of plutonium have been developed with different objectives (e.g., whereas one approach may be best at reducing the risk of environmental and human health impacts, another may be better for reducing the risk of proliferation, and yet another for extracting economic value from the plutonium). The tradeoffs among different approaches cannot be analyzed reliably until more research has been completed.

The more advanced technological approaches have significant uncertainties about when they might be available for full-scale development (30), how effective they might be, the development effort involved, what other impacts might result, what nontechnical barriers may arise, and what benefits they might offer (18). The costs to implement most of these technologies are not well known at present (30). Plutonium burning as U.S. mixed-oxide fuel[18] in conventional light-water reactors (LWRs), abandoned in the United States in the 1970s, is probably the best cost option (18,59). Costs of some of the fission options have been estimated by their proponents, although a detailed comparison of costs and assumptions has not been made (30).

Therefore, the following discussion of plutonium processing should be interpreted as a very early indication of how to approach the question

[18] MOX is made by mixing the oxides of plutonium and uranium, and forming the product into conventional reactor fuel assemblies.

Box 4-C—Plutonium Processing

Plutonium processing is used in this report to define a myriad of manufacturing steps that may be employed to change the form, configuration, content, and chemical or radiological state of the material. The purpose of these changes could be to make plutonium usable as reactor fuel, to make it more stable for storage, to prepare it for disposal, or to alter its radiological state so as to eliminate long-lived radionuclides. The steps may include chemical, thermal, mechanical, and radiological (neutronic) processes. One near-term proposal for processing involves making mixed-oxide (plutonium and uranium) fuel and then using it in a nuclear reactor. Although this technology is available in some other countries, no facilities for carrying out these processes exist in the United States.

Many steps are required to make mixed-oxide (MOX) reactor fuel from plutonium pits. First, the plutonium metal must be removed from other associated materials—by chemical or mechanical techniques—then the metal can be purified, probably by a chemical solution process. Next, pure plutonium would be converted to plutonium oxide by calcination and finely pulverized to improve its reactivity. The plutonium oxide would then be blended with depleted uranium oxide or natural uranium. The uranium oxide would have been derived from enrichment plant residue by using a chemical step to convert uranium hexafluoride to uranium oxide and then finely pulverizing it. The mixture of uranium and plutonium oxides would be pelletized, sintered, and loaded into fuel tubes to be used in a reactor as MOX fuel. Each step in this process must be carefully controlled to prevent releases, protect workers, and ensure safety. Each step also produces some waste or scrap. Some of the waste may be recycled, and some would be treated as transuranic waste because of its reduced plutonium content. Some waste may be mixed with hazardous and toxic chemicals or other materials.

The waste generated by MOX fuel fabrication could contain about 1 percent of the plutonium input and 5 percent of the uranium input to the process, but the quantity of waste product would be significantly higher because it would be mixed with other materials, much of it hazardous waste itself. The experience with plutonium processing at the DOE Rocky Flats Plant is a case in point, in which huge quantities of residue and waste still exist without a good disposal solution. Whatever waste is produced will require appropriate systems for storage, treatment, and disposal as well.

It should also be noted that after MOX fuel is made, the remainder of the fuel cycle, mainly within a nuclear power reactor system, also produces waste that must be properly controlled and handled. Finally, the disposal of spent fuel after irradiation in the reactor presents another waste disposal problem. As discussed elsewhere, all spent fuel from standard U.S. nuclear reactors is stored temporarily at reactor sites awaiting an acceptable solution to its ultimate disposal. Spent MOX fuel would be subject to similar constraints.

The above is an example of just one plutonium processing option with its related waste generation and disposal issues. This report discusses many other processing approaches—making plutonium oxide forms to enhance storability; mixing plutonium with other wastes and vitrifying the mixture to enhance disposability; or transforming plutonium radioactively in advanced reactors or accelerators to change a large percentage of it into other, shorter-lived radionuclides. These processes would also include the generation of wastes and thus must be properly managed to protect human health and the environment.

SOURCE: Office of Technology Assessment, 1993.

of ultimate disposition of this material. Whatever path is pursued, it will be necessary to carefully investigate technical feasibility, impacts on health and the environment, ability to meet ultimate disposition goals, and possible economic benefits (18). Although varying amounts of technical information are already available for some options, such as vitrification or use as MOX fuel, these options generally have not been evaluated and compared on their merits *specifically* as options for processing surplus military plutonium in the United States (59). No best approach can be selected today with confidence. After some initial evaluation, however, a few approaches could be researched and their merits identified. If clear policy goals have been adopted, then the most technologically developed approaches could be compared more readily and an optimum one selected.

Use of Plutonium as Mixed-Oxide Fuel in Light-Water Reactors

Various options that call for plutonium to be used as a fuel in nuclear reactors have been proposed. These options are based on incremental changes in currently available, working technology. One option would involve incorporating plutonium in mixed-oxide fuel to substitute for some of the conventional low-enriched uranium fuel used in commercial LWRs. Proponents point to the electricity generation potential of this disposal option as an economic advantage, whereas opponents claim that MOX fuel cannot compete with ordinary LEU fuel economically (16). Another option would use the plutonium incorporated into MOX fuel in dedicated reactors that could be built on a Federal site, primarily to convert plutonium into more proliferation-resistant spent fuel elements and possibly to produce some electrical power as well, whose sale could offset some costs of the project. See box 4-C for a description of the facilities and steps required for a MOX-fueled reactor approach.

Some experts claim that the use of plutonium as MOX fuel in nuclear reactors is advantageous because after irradiation, the fuel would be poisoned with very toxic fission products that make plutonium recovery difficult for any group without reprocessing facilities (5). It is technically straightforward to substitute MOX fuel for about one-third of the LEU fuel used in conventional light-water reactors such as those in the United States. However, the use of MOX fuel in *existing* LWRs in the United States is viewed by many as detrimental to verification and proliferation resistance because the practice would distribute plutonium widely in the commercial sector (3). An alternative would be to have fewer specially designed reactors that could use 100 percent MOX fuel loadings in order to minimize physical distribution of the plutonium and thus enhance both verification (by on-site inspection) and proliferation resistance. However, utilities are uncomfortable with the prospect of using plutonium as fuel for civilian power reactors. They believe that public opposition may constrain such practices and that the regulatory process would be long and difficult (3,67).

Although the notion of recovering value from weapons plutonium by converting it to MOX fuel is attractive to some, there are drawbacks to this option. No MOX fuel fabrication facilities currently exist in the United States.[19] There are, however, MOX facilities in other countries. A large MOX facility, owned by Belgonucleaire, is located in Dessel, Belgium. Its startup and status are currently being debated in that country (15). The Siemens company built a facility in Germany that was designed to convert plutonium into MOX fuel, but operations have been delayed indefinitely. In Russia, the Ministry of Atomic Energy (MINATOM) plans to continue to reprocess spent power reactor fuel. MINATOM may also use separated civilian plutonium as fuel for its fast neutron or other reactors. Construction of an industrial-sized facility at Chelyabinsk 65, intended to manufacture fuel for three BN-800 nuclear reactors to be built at the site, was suspended. MINATOM would probably like to find outside financing to complete the MOX facility to manufacture MOX fuel for other existing reactors or even for future breeder reactors (17). However, there is also opposition to nuclear power expansion plans in Russia based on economic and environmental concerns.

If the United States built special dedicated LWRs, plutonium in MOX form might be used

[19] Two MOX fuel fabrication facilities were constructed at Hanford to supply the now-canceled Clinch River Breeder Reactor. The facility was never operated, and it is unlikely that it could be reopened to comply with modern safety and environmental standards.

and converted to spent fuel. Plutonium would remain in the spent fuel but be mixed with highly radioactive fission products that would make it significantly less of a diversion risk (3). One analysis estimated that six 1,000-megawatt reactors operating for a decade with full core loadings would convert about 50 tons of plutonium into spent fuel (3). If the same amount of plutonium were used as fuel in conventional reactors with one-third reactor core loadings, the number of reactors required would increase threefold. This could add to diversion risks (3). Dedicated reactors could be built as specially adapted, safeguarded, and secured for this purpose and probably located at a Weapons Complex site (3). However, a potential drawback, according to some observers, is that construction and operation of any plutonium fueled reactors might encourage the United States to adopt a permanent plutonium fuel cycle that could increase the risks of plutonium diversion and proliferation (15,16).

A few, very preliminary economic analyses have been done of the use of weapons plutonium as MOX fuel in civilian power reactors. Some of these, while emphasizing the proliferation, verification, security, and monitoring aspects of plutonium disposition, conclude that there are no economic benefits in the use of weapons plutonium as fuel in commercial reactors, even if the plutonium itself is "free." At the current relatively low price of uranium, it would cost more to convert plutonium into MOX fuel and substitute it for LEU fuel in conventional LWRs (67). One estimate is that the fabrication cost of combining plutonium—which is more hazardous to work with than uranium—into MOX fuel is "at least" twice the cost of LEU[20] fabrication (3). However, others emphasize the inherent value contained in weapons plutonium and see it as an asset to be exploited. A related viewpoint is that economic cost-benefit arguments for any option are unlikely to be key criteria when measured against the importance of making plutonium less usable for weapons (5). Finally, according to another analysis, the cheapest and quickest way to get surplus plutonium into a more proliferation-resistant or long-term disposal form would be by some direct disposal option (17).

Other nuclear experts have noted that even if the primary goal is to convert surplus weapons plutonium into a proliferation-resistant waste, then a method such as burning in a MOX reactor (which would also generate some electricity) may be attractive. Studies of possible MOX fuel use have been performed by two utility industry groups (the Electric Power Research Institute and the Edison Electric Institute). These studies conclude that the once-through option[21] has merit (49,60). These studies also support the construction of a dedicated MOX-fueled reactor facility on a Federal site. Such an approach would avoid a major change in U.S. commercial regulatory policies and would enable the existing security and other infrastructure to be used.

Over the next 10 to 20 years, MOX-fueled LWRs may have the technical potential to dispose of large quantities of plutonium with partial core loadings of MOX fuel (5). This would require that facilities be built to convert plutonium to plutonium oxide, mix it with natural or depleted uranium oxide, and then manufacture MOX fuel.

A future problem in need of attention is that spent MOX fuel would eventually require disposal. Indeed, all schemes that call for use of plutonium in reactors produce spent fuel. There are no operable, long-term disposal facilities for spent fuel from commercial reactors. The outlook for geologic repositories for spent fuel is uncertain. Investigations of a possible repository site in Nevada have encountered serious delays and public opposition, and are unlikely to be completed soon.

---

[20] The term "low-enriched uranium," or LEU, denotes fuel for conventional power reactors that contains 3 to 5 percent uranium-235.

[21] "Once-through burn" refers to the use of MOX in nuclear reactors as a means to convert the weapons-grade plutonium in MOX to the more proliferation-resistant reactor-grade plutonium. No recycling of the weapons plutonium embedded in spent nuclear elements is involved.

Developing the facilities and transportation for using plutonium as a fuel in civilian nuclear power reactors also poses special problems for nuclear proliferation and security. The environmental, safety, and health impacts of the processing of plutonium through MOX reactor fuel require updated investigations.

A final problem facing any proposed use of MOX fuel in commercial reactors is current U.S. practice of not recycling commercial plutonium. Such a policy was established in the 1970s after a long debate about commercial plutonium reprocessing and use in breeder reactors. A Generic Environmental Impact Statement on Mixed Oxide (GESMO) was the focus of this debate. The GESMO project was terminated in 1979 when the Carter administration announced the policy not to pursue plutonium recycling. Although this does not currently prohibit the use of MOX-fueled reactors, a new Environmental Impact Statement would be required, and many believe that the 1970s debate would be rekindled (32,60).

Other Plutonium Fission Options

Several other fission options have been proposed for plutonium processing. One approach envisions the use of "fast" reactors with a metal fuel cycle. Under some conditions, fast neutron reactors may be able to fission plutonium more quickly than light-water reactors. Several countries, including the United States, are developing fast neutron reactors—usually as "breeder" reactors that produce, rather than consume, plutonium. None is available today as a proven means of plutonium disposition.

A recent study prepared for DOE's Plutonium Disposition Task Force (48) appears to describe salient features of most of the known approaches with current data; this study concluded that use of excess weapons plutonium in fission reactors could address multiple goals. Certain options were compared on the basis of proliferation resistance, environmental protection, and power generation to offset operating costs (48). Fourteen different fission options were considered. The study estimated that the time required to deploy them ranged from 5 to 25 years, if the resources to support such development were available. The study estimated the remaining development costs of these options to range from \$0.1 to \$10.0 billion each and concluded that, when developed, most options would be able to produce sufficient power for sale to offset substantial portions of the operating costs.

It is clear that this study began with the notion that weapons plutonium is an asset. The options were selected and compared with the primary goal of obtaining a return on this asset while meeting an additional goal of making the material resistant to diversion for future weapons use. Recommendations made by this study are that some of the options appear quite promising and should be analyzed in greater detail. The advanced light-water reactor option with full MOX core appeared to be the best for relatively early deployment, and advanced concepts such as the accelerator-based converter had the best potential for achieving the greatest degree of plutonium transformations into more benign elements and shorter-lived radionuclides.

A number of concepts featuring advanced reactor or converter designs have been proposed with plutonium disposition as a primary objective. They all involve nuclear fission reactions in a device that focuses on long-lived radioisotopes such as plutonium and attempts to produce such reactions more efficiently than current reactor designs.

The descriptions in boxes 4-D, 4-E, and 4-F illustrate some of the technologies examined. Each represents an example of an advanced technological approach. The first is the advanced liquid metal reactor—a concept that, according to Omberg and Walter (48), would take about 10 to 15 years to deploy. This estimate is regarded by certain experts as highly optimistic. The second is the modular high-temperature gas reactor, a concept with a 10- to 20-year development time, and the third is the accelerator-based converter, a

Box 4-D—Plutonium Transformation Concept 1: Advanced Liquid Metal Reactor/Integral Fast Reactor System

The advanced liquid metal reactor/integral fast reactor (ALMR/IFR) has been proposed as a plutonium disposition option. It was originally designed as a fast breeder reactor for electricity generation (producing more plutonium than is consumed).

The ALMR design could be modified to consume plutonium and other transuranic actinides instead of producing them. This feature was promoted as a means to eliminate such actinides in spent fuel from conventional U.S. light-water reactors. It would still require plutonium reprocessing, and many burning/reprocessing cycles would be required to significantly reduce the actinide inventory in spent fuel. This proposal is currently being evaluated by the National Academy of Sciences Panel on Separations Technology and Transmutation Systems (STATS panel).

With a new interest in disposal of surplus military plutonium, ALMR designers have suggested the possible use of their design. However, the concept of plutonium transformation using fast reactors appears to have some limitations. To consume plutonium in a fast reactor requires significant design changes from the original LMR that was intended to produce plutonium. It could also be expensive: the required reprocessing could multiply the total volume of radioactive waste by 10, thereby driving up costs (7).

The concept also envisions reprocessing, to separate fission products in spent fuel, and subsequent recycling of the remaining plutonium. The licensing process would likely be difficult and contentious both for the ALMR facilities and their associated reprocessing facilities (45). Reprocessing would be either a standard chemical separation process or a pyrochemical process if one was sufficiently developed. Aqueous waste from the process would contain transplutonium actinides including neptunium and residual plutonium, although another process under development at Argonne National Laboratories can recover better than 99.99 percent of all actinides, leaving only fission products in the waste solution (6). Fuel fabrication with recycled plutonium (after the first cycle with pure weapons-grade plutonium) would have to be done remotely in a hot cell because of gamma-emitting actinides (52).

If it operates according to present designs this option would eliminate most transuranic actinides, including plutonium, while generating high-level waste. That waste would require a repository, the future availability of which is unknown.

Deploying ALMRs solely for burning weapons plutonium would be difficult to implement because only a small amount of plutonium may be made available from weapons dismantlement. Proponents usually tie this concept to a national decision to turn to a plutonium breeding/recycling energy program. Moreover, as a strategy to eliminate actinides including plutonium contained in spent fuel, this would be very slow compared to many other direct disposal strategies such as vitrification. To reach a tenfold reduction in the inventory of actinides accumulated in U.S. spent nuclear fuel (equivalent to burning 90 percent) was estimated to require more than 100 years (45).

SOURCE: Office of Technology Assessment, 1993.

concept that would take 20 to 25 years to develop according to the study. Figure 4-4 illustrates the major steps involved in these three alternative reactor approaches.

In a general sense, all these concepts attempt to convert one atomic species or radionuclide to another. A significant percentage of a long-lived radioisotope (e.g., plutonium-239) is converted to either shorter-lived radioisotopes or stable isotopes by reaction with neutrons produced in a nuclear reactor or neutrons created by bombardment of a metal target in an accelerator.

The three concepts discussed in boxes 4-D through 4-F are merely illustrative of a larger

Box 4-E—Plutonium Transformation Concept 2: The High-Temperature Gas-Cooled Reactor

The high-temperature gas-cooled reactor (HTGR) concept has been under development for other purposes for a long time. Its predecessor was the gas-cooled reactor designed by General Atomics and operated at Peach Bottom, Pennsylvania in the 1970s. More recently, the modular HTGR (MHTGR) concept was proposed as the basis of a new generation of reactors; it was also a possible choice for the new production reactor to produce tritium for weapons. Proponents of this concept claim that the reactor could act as a plutonium burner, converting a large percentage of weapons plutonium-239 to plutonium-241 and plutonium-242.

The MHTGR reactor uses fuel particles coated with ceramic materials that allow the long-term, high-temperature operation desirable for efficient energy production. The neutrons in the core are moderated by graphite, and the reactor is cooled with helium gas. Designers claim that reactor safety is based on inherent characteristics, physical principles, and passive design features, rather than on active engineered systems, operator actions, evacuation or sheltering, or even reactor vessel structural integrity. Core melting is not supposed to occur even with a loss of coolant accident because of the refractory nature of the fuel. The reactor is contained underground for added safety.

MHTGR designers have studied several options for "burning" weapons-grade plutonium. In one concept, more than 90 percent of plutonium-239 is consumed. The spent fuel discharged after 2 years contains roughly 40 percent plutonium-241. Although plutonium-241 is fissile, its half-life is only 14.7 years, much less than plutonium-239. In one reference design, 50 metric tons of weapons plutonium could be irradiated in six 450-megawatt (thermal) plutonium-fueled MHTGR modules over 40 years. The spent fuel packages would have some fissile materials in them but would also be contaminated with nonfissile actinides and long-lived fission products.

Developers of the MHTGR concept point out several weaknesses. This would be a "first of a kind" reactor with concomitant high costs. There would have to be a program to develop and verify performance of the fuel and to develop fuel manufacturing capabilities. Also, the experiential base for this reactor concept is weak. The concept has also been proposed by the developers for application in Russia. A further claim by the developers is that it could be used for both tritium production and plutonium destruction.

SOURCE: Combustion Engineering/General Atomics.

number of possible approaches. One of these three, the accelerator-based converter (ABC), which involves the partitioning and recycling of long-lived fission products and actinides, is claimed to be capable of destroying essentially all the plutonium—a process termed "annihilation" by Omberg and Walter (48). The other two are said to be somewhat less thorough than the ABC at completely fissioning plutonium and result in fission of some part of the original plutonium. All of these concepts involve considerable uncertainty, and much more work would be required to determine their feasibility. Each requires the development of various technologies and systems beyond the basic reactor or converter itself, such as systems for fuel fabrication and preparation and for waste treatment.

All technologies potentially capable of extensive conversion of plutonium will require substantial investments in development to move them closer to viability. The development time and effort required for most of these concepts have not been thoroughly investigated. Many claims have been made by proponents of certain systems, but they have not been compared on an impartial basis. A National Academy of Sciences panel is studying the transmutation of actinides in

Box 4-F—Plutonium Transformation Concept 3: An Accelerator-Based Converter, the Los Alamos Concept

Los Alamos National Laboratory (LANL) has been examining the potential for using accelerators to bombard targets in order to produce sources of neutrons to serve a variety of functions, including production of tritium, destruction of actinides and high-level radioactive wastes, and destruction of weapons-grade plutonium (94 percent plutonium-239 and 6 percent plutonium-240) by converting it to plutonium-242 and fission products. In this converter, a high-current proton accelerator would bombard a heavy-metal target, producing an intense source of thermal neutrons that interact with the weapons-grade plutonium. The plutonium would be fed continuously in a carrier medium, and the discharge would be separated and remaining plutonium recycled.

Advocates of this concept claim that advances in accelerator design stemming from the Strategic Defense Initiative, combined with experience in developing intense central neutron sources at Los Alamos, have brought this concept closer to reality. One virtue of the concept is that it does not involve use of a critical nuclear reactor. Although a subcritical reactor may be theoretically safer than a reactor that requires criticality to operate, the safety of a subcritical reactor with an intense neutron source has not been thoroughly evaluated. If the concept works, rapid destruction of plutonium-239 to very low concentrations may occur.

The LANL accelerator-based transmutation concept appears attractive for destruction of weapons plutonium. However, questions remain about its technical feasibility. The concept appears to be technically daunting, requiring the state-of-the-art development of three technologies—accelerator; subcritical reactor; and on-line, continuous reprocessing system. It will also produce radioactive activation products and wastes. Furthermore, the LANL concept involves on-line continuous processing and recycling of actinides and long-lived fission products, which would likely present significant health and safety hazards.

SOURCE: Office of Technology Assessment, 1993.

nuclear waste.[22] Its results, due in 1994, could have applicability to the disposition of surplus weapons plutonium.

The Omberg and Walter study (48) reviews the time and effort for developing various plutonium fissioning concepts and indicates that if several approaches are pursued simultaneously, a multi-billion-dollar program spanning a few decades would be required before actual, full-scale systems could be tested and proven. A number of skeptics in the scientific and engineering community doubt promotional analyses claiming that certain options can both destroy plutonium and yield economic returns. Any program to develop these technologies should be based first on a clear overall national policy regarding the disposition of weapons material and second on an impartial, high-level, Federal Government evaluation of costs, benefits, and uncertainties.

It was not possible for OTA to conduct an independent analysis of the merits of various plutonium fission options. However, based on its general analyses, OTA concludes that it will probably be necessary to choose among options before they have been fully studied and developed. Unless a national policy is articulated in a timely manner, large amounts of time and money could be spent on options that turn out to be contrary to future U.S. policy.

One key policy choice is whether or not plutonium should have a place in international commerce. If the answer is yes, the United States will have to develop the means to manage plutonium over the long term for possible useful economic purposes. If the answer is no, the

[22] The panel on Separations Technology and Transmutation Systems (STATS) was organized in late 1991 and has investigated advanced reactor and converter technologies that could be applied to high-level radioactive waste management problems.

Figure 4-4—Selected Advanced Reactor/Converter Options for Plutonium Disposition

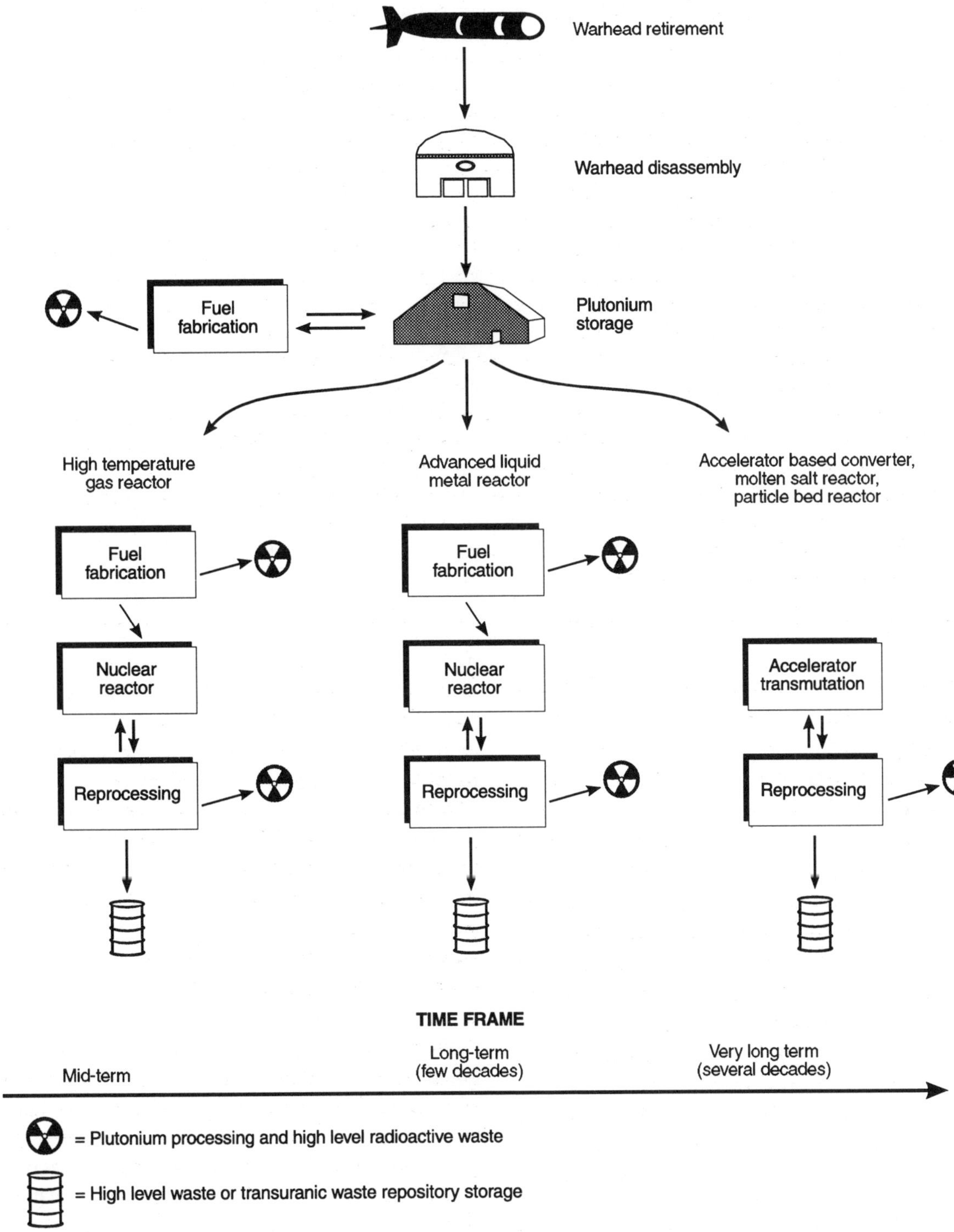

SOURCE: Office of Technology Assessment, 1993.

United States must find an acceptable means of processing plutonium via a reactor or directly disposing of it. Another key policy decision is whether plutonium should be put into a less weapons-usable form as quickly as possible. If this is an overriding goal, then technologies available in the near term would be favored over those requiring long and uncertain development.

Proponents of plutonium as waste and other experts conclude that the primary goal is to convert plutonium quickly to a form that is most difficult to extract and reuse in weapons. If this goal is accepted, then research could be directed to determine whether disposal of plutonium as waste is technically feasible and can be accomplished safely at reasonable cost. If conversion into waste is a goal, the best solution may be disposal with as little processing as possible. The option of plutonium disposal without conversion might be desirable because the infrastructure for plutonium utilization is not in place in the United States and there is significant public concern about its use (5). Moreover, a key reason for the U.S. abandoning the development of a plutonium infrastructure in the 1970s was concern that it would encourage worldwide plutonium proliferation.

Criteria for Treating Plutonium as Waste

The efficacy of treating plutonium as a waste may be gauged by the following criteria:

- Security. The treatment, storage, and disposal of plutonium must be such that the difficulty of plutonium reextraction from the waste is high.
- Accident. The risk of catastrophic accidents must be evaluated.
- Health and safety. Processing plutonium as a waste form involves consideration of, and protection against, health and safety risks to workers and the public.
- Long-term management. Because of its long half-life, plutonium must be isolated from the human environment for extremely long periods, and waste treatment must be compatible with long-term management.
- Cost. Some nuclear experts believe that the security benefits of converting plutonium into a waste form that is proliferation resistant far exceed any potential economic benefits from its use. The uncertainties associated with most proposed approaches make cost evaluations very difficult. However, nonproliferation benefits and the value of doing something quickly must also be weighed.

A number of waste disposal options for plutonium have received some attention, including:

- disposal in a geologic repository,
- sub-seabed disposal,
- detonation of warheads underground to fix plutonium in molten rock, and
- disposal in outer space.

DISPOSAL IN A GEOLOGIC REPOSITORY

Plutonium could be disposed of as a waste in a geologic repository. It could be disposed of directly after being packaged in special containers or immobilized in a vitrified form prior to disposal. Criticality requirements, however, must be developed and accepted.

Direct plutonium disposal in a repository also requires consideration of other factors. Plutonium would have to be packaged in small quantities and in special containers to prevent accidents. Increased criticality concerns and the potential for recovery of plutonium from the repository may open up new questions regarding repository licensing. A serious argument against such direct underground disposal is that the plutonium could be recovered easily in the future and, if not recovered, could pose a significant risk of contamination unless immobilized in a matrix.

If direct disposal of plutonium were unacceptable, the next approach might be to encapsulate it in a form that could potentially retard its dispersal into the environment. Encapsulation technology could also make it difficult and costly to recover the plutonium for reuse, compared with new plutonium production.[23] One option is to vitrify plutonium without adding any products except glass. Experts at DOE's Savannah River Site have been investigating methods to produce vitrified glass containing a small percentage of weapons plutonium. A second option is to mix plutonium with high-level waste or poisons prior to vitrification. Most experts agree that if appropriate "poisons" or other products are added to plutonium, it can, in theory, be made as proliferation resistant as spent fuel.

Encapsulation technology has been examined extensively for the high-level waste resulting from plutonium production (most of the waste is now in large tanks at Hanford and Savannah River). A number of different materials with a wide range of properties for encapsulation have been considered (including different forms of glass, ceramics, and cement-related materials, along with various metal coatings).

These materials possess varying properties in relation to the isolation of high-level radioactive waste. Most of them have not been thoroughly evaluated, manufacturing technologies are not fully developed, and knowledge of their applicability to weapons plutonium is limited. In 1982, DOE chose borosilicate glass as the waste form at the Savannah River Site partly because the manufacturing technology for glass was far more advanced than that for other proposed waste forms.

Because glassification of radioactive waste is an available technology (at least in countries such as France and the United Kingdom, although not quite operational in the United States), encapsulation of plutonium in glass could, in theory, provide a relatively short route to disposition of plutonium as a waste. Two plants for the vitrification of high-level waste from reprocessing have been built in the United States. One is at West Valley, New York (the West Valley Demonstration Plant); the other, the Defense Waste Processing Facility, is at the Savannah River Site in South Carolina. Both are DOE facilities. Even though these facilities are nearing startup, they have suffered long development or implementation delays. Glassification is the most near term of any technology, but the remaining engineering and testing required should not be underestimated (66).

Although borosilicate glass has been investigated more extensively, other waste forms may possess better isolation properties for actinides—an important factor in light of the 24,000-year half-life of plutonium-239. The use of these other waste forms for plutonium has the disadvantage that much more research and development are required, and thus the relative costs and benefits are unknown. However, it may be useful to explore alternatives to borosilicate glass for plutonium vitrification, some of which could be more desirable for reducing long-term releases.

Plutonium pits from warheads would have to undergo some processing before being vitrified. Plutonium metal is too chemically reactive, pyrophoric, and insoluble in glass for vitrification. Suitable forms of plutonium for vitrification include plutonium dioxide (powder or particulate form) and plutonium nitrate. It may be necessary to mix plutonium oxide or nitrate with other materials prior to or during vitrification. Calcined materials that could be mixed with plutonium for vitrification already exist at the Idaho National Engineering Laboratory and the Hanford Plant.

In May 1993, the Westinghouse Savannah River Corp. issued a draft report on vitrification of plutonium (79). The study provides technical

---

[23] Plutonium production involves irradiation of uranium fuel in a nuclear reactor, followed by chemical separation of plutonium from the remaining uranium and fission products. Because the fission products are very radioactive and toxic, chemical separation requires elaborate facilities with extensive shielding, such as the canyon facilities at Hanford and the Savannah River Site. Recovering plutonium that has been encapsulated in glass along with high-level waste would be similar to the chemical separation used to produce new plutonium—and therefore would require access to elaborate and extensively shielded facilities. This is something a large nation might support but a terrorist group might not be able to.

information about several vitrification options—some using existing facilities with modification and some requiring new facilities. The report concludes that the most straightforward option with only slight modification of existing facilities would require almost 10 years before beginning operations. Some rough costs are also given in this study. The least costly option was seen as vitrification without addition of a radiation source, and vitrification in a modified reprocessing canyon with added radioactivity would be a high-cost option. Total costs for vitrifying 50 tons of plutonium range from $0.7 to $1.6 billion. Finally, the report notes that research and development for all options is still needed on criticality safety, defining physical and chemical properties of the glasses, and developing and demonstrating performance of processes and waste form.

More detailed, quantitative, environmental, safety, and cost analyses are required to fully assess all options for using either existing or planned high-level waste vitrification plants or, possibly, a new plant built exclusively to vitrify plutonium. Worker health and safety considerations would require particular attention to radiation protection measures, especially if fission products are combined with plutonium and vitrified. Different options would imply varying storage times for the plutonium from dismantled warheads because of different startup times for facilities. The composition of the glass is also important for its long-term isolation properties, which will be crucial in protecting the environment from eventual contamination after the disposal of vitrified plutonium. Also, depending on the product (plutonium and glass alone, or mixed with poisons and wastes), the difficulty of future recovery of plutonium may vary considerably. Although many countries do not have the technology to retrieve plutonium vitrified with high-level waste, certain nuclear countries such as the United States and Russia do. In terms of costs, one must evaluate the economics of plutonium recovery from glass relative to the production of new plutonium from reactors and reprocessing plants.

In summary, the direct disposal of plutonium as a waste—like the option of disposal as spent fuel after plutonium irradiation—would depend on the availability of a radioactive waste repository. No such repository is now available in the United States nor is one likely to become available in the near future. A minimum of a few decades will probably be required before a geologic repository for high-level commercial spent fuel can be opened, but so many technical and political setbacks have been encountered during the past decade that it is difficult to make realistic predictions.

Two approaches are most likely in considering the prospects for disposal of weapons plutonium in a repository—one is to plan for indefinite storage of whatever the short-term form is (from pits to glass logs); the second is to highlight the need to develop long-term solutions for this problem, as well as the problem of disposal of other defense and civilian radioactive wastes.

SUB-SEABED DISPOSAL

Another option that some have advocated is sub-seabed disposal of plutonium either directly or as glass logs with waste (57). Significant investigations have been done in the past on sub-seabed disposal of spent fuel from commercial reactors (64). These investigations were suspended several years ago, but some researchers have suggested that it may be appropriate to study this option for weapons plutonium disposal. Here again, more analysis is needed to determine the costs and benefits, and public and international acceptance may be a formidable obstacle.

UNDERGROUND DETONATION

The option of detonating a nuclear bomb underground as a means of fusing plutonium into the surrounding rock was suggested by a group of Russian scientists. Some believe the verifiability of this option to be good (5). Costs of this nuclear explosion glassification process might also be low, but no good analysis is available. Irreversibility is problematic because of the possibility of

ROBERT DEL TREDICI

The Defense Waste Processing Facility (not yet operational) at the Savannah River Plant. Some have suggested using this facility for vitrification of plutonium mixed with high-level radioactive waste from past operations.

recovering the fused rock and leaching out plutonium (5). The safety and environmental impacts of this option have not been evaluated to any degree, and these concerns have blocked support for serious analysis.

Political and public acceptance would probably be extremely difficult to obtain in the United States, if not worldwide, and recent decisions about stopping nuclear testing in the United States and elsewhere could be affected by a serious consideration of this option. Thus some consider this an "option of last resort" (5).

DISPOSAL IN SPACE

The option of deep space disposal of plutonium could offer irreversibility, proliferation resistance, and verifiability. Concerns about the safety of such a project center on the possibility of accidents during launch, with the potential for plutonium dispersion over large areas (5). Costs, although currently difficult to estimate, may be *much* higher than for other options such as geologic disposal, although this could be subject to reevaluation. Very little analysis has been done on the space disposal option, and almost no attention has been given to it in the past 10 years. Most experts have relegated consideration of space disposal to the bottom of the list.

PLUTONIUM DISPOSITION—CONCLUSION

The discussion above has presented a variety of approaches for the disposition of plutonium in the United States from retired and dismantled nuclear weapons. The following concluding points summarize OTA's analyses of the available data, with reference to the technical and political factors in the United States. Some aspects may be applicable to other countries such as Russia, but different conditions can also result in very different conclusions. Plutonium disposition is considered by many to be one of the most difficult problems faced by those who will manage materials from retired nuclear weapons. Not only is it a difficult problem, but it also must be considered in the wider geopolitical context of security, human health and safety, and the environment.

- Storage is a necessary first step, regardless of which approach is selected for the ultimate disposition of plutonium. The questions regarding storage are, How long? In what form? What kind of facility? Where? Decisions about ultimate disposition are unlikely to be made soon, but even if they are, significant portions of the plutonium stockpile will be stored for decades. Thus, it makes sense to move toward a safe, secure, state-of-the art storage facility rather than rely on politically sensitive temporary facilities such as those at Pantex, with risky periodic lifetime extensions.
- The use of weapons surplus plutonium as fuel for U.S. commercial reactors is unlikely in the near term because of economic factors and the concerns of U.S. utilities about regulatory constraints and public opposition. Further, U.S. policies that discourage commercial plutonium use because of proliferation concerns would need to be reevaluated.

- The use of a modified light-water reactor system for disposition of plutonium as mixed-oxide fuel at a dedicated government facility is probably a viable near-term approach if proper attention is given to worker and public health and safety, environmental protection, and public involvement.
- It may be possible to immobilize plutonium directly into some waste form such as vitrified glass, with or without high-level waste fission products. This approach could offer proliferation resistance. A rigorous analysis of the costs and benefits of this approach, compared with reactor approaches (e.g., dedicated reactors with 100 percent MOX fuel loading) that involve subsequent handling of the spent fuel, would be very useful. Here again, health, safety, and environmental protection would need adequate attention.
- Decisions about the fate of plutonium from U.S. weapons should be made with consideration of Russia and other nations that may be planning to use plutonium in reactors. Policy goals should be stated clearly. If the United States wishes to reduce the world stockpile of plutonium that is easily available for weapons, it should take actions to discourage future production, control existing materials, and make them unusable for weapons.
- It is all but impossible to fission plutonium completely (and thus "destroy" it), but future technological developments may have the ability to convert it to different radionuclides more effectively than any existing system. Research into advanced reactors and accelerators would be costly and require long development times (decades), so any program should focus on specific goals. Research into space disposal or other unconventional options may merit limited support if they can be justified on the same basis.

DISPOSITION OF HIGHLY ENRICHED URANIUM

Substantial quantities of highly enriched uranium will result, by the end of this decade, from the dismantlement of retired weapons. The U.S. government has made no decisions regarding whether or when weapons-grade HEU will be available outside DOE programs. The technology required to use HEU in commercial or other reactors, after blending it down to LEU, is considered simple by many. The logic is that it will be easy to shunt weapons-grade uranium into the world's already established uranium-based nuclear power industry. Therefore, the interest in pursuing research into innovative HEU disposition options is sparse.

Significant attention, however, is focused on the purchase of surplus HEU from Russia and the consequent use of that material as fuel by the U.S. commercial nuclear power industry (see chapter 6). U.S. purchases of HEU would provide hard Western currencies that Russia desperately needs to bolster its economy and would guarantee that some Russian HEU will not be used for making new nuclear warheads.

However, OTA's analysis indicates that some problems must be addressed before a program to utilize warhead HEU can be implemented. More extensive investigation is needed of the following: the dilution and conversion of warhead HEU to the LEU used in commercial power reactors; the testing and operation of conversion facilities; interim storage prior to conversion; assurance of adequate safety, security, and verification in processing and transport; the impact of weapons surplus uranium on the already depressed U.S. and worldwide uranium industry; and the uranium dumping suit brought against the former Soviet Union by the U.S. Uranium Miners Union and others. It will also be important to develop clear national policies about what to do with U.S. military uranium in light of future security needs. These considerations will influence any decision on HEU disposition that may be made in the future and should be part of the present planning process even if no decision beyond storage is being considered at present.

DOE is reluctant to quickly convert U.S. weapons HEU for other purposes. Some time will also be required to bring any HEU processing operation on-line and deal with possible disruptions in the uranium market. It appears likely that HEU (like plutonium) will have to be stored in a safe, secure manner for the immediate future.

The United States has stopped production of HEU and is not planning to make any additional HEU, at least in the near future. In 1992 the Bush administration stated that U.S. policy was not to make any more HEU for nuclear weapons and that DOE had actually ceased producing HEU specifically for weapons in 1964 (22). This announcement formalized what circumstances had already dictated. The production of nuclear weapons plutonium effectively ceased after 1988 because of safety and environmental problems at DOE reactors and weapons plants (62). However, DOE continued to produce HEU until 1992 for the nuclear Navy, research reactors, and defense production reactors. In addition, the U.S. decision to cease all HEU production was the recommendation of a high-level task force formed in 1991 to examine HEU options in light of the large amounts of HEU expected from dismantled warheads (62).

It is not certain what fraction (if any) of the HEU coming from retired U.S. warheads will be converted to civilian fuel,[24] as opposed to being kept for military purposes such as fuel for naval reactors (which presumably could be modified to use the slightly lower enrichments) or to make new nuclear weapons. The possibility of converting a portion of U.S. military HEU for sale in the commercial LEU industry is being considered seriously by some. In its report on the National Defense Authorization Act of FY 1993 (Public Law No. 102-484), the House Committee on Armed Services requested a cost-benefit analysis of blending surplus HEU with LEU and uranium scrap for use as commercial reactor fuel (14,21).

Some U.S. utilities would also like to see U.S. military HEU blended to LEU and made available on the market as fuel for civilian power reactors in a manner similar to current plans for Russian military HEU. The first U.S. military uranium that may be converted to civilian commercial reactor fuel would probably be HEU that is in DOE's inventory but not in warheads. Generating LEU fuel by blending down HEU, instead of mining more uranium ore and enriching it, is environmentally advantageous because it would avoid the land contamination associated with mining as well as the energy expenditure associated with uranium enrichment.

At present, there is no apparent effort in the United States to make available any HEU recovered from dismantling warheads (35). Nevertheless, the United States may come under some pressure to show reciprocity by converting its HEU to other uses, if it can be assured that the Russians are converting their military uranium to civilian purposes (as required by the pending Russian HEU agreement). However, the possible demand for reciprocity in nuclear warhead dismantlement has not received official attention (38,54). Most Russian officials have expressed more interest in the economic value of HEU than in its security value (38). The major pressure so far for reciprocity has been from other groups and other nations—particularly related to renewal discussions of the Non-Proliferation Treaty coming up in 1995. Some believe that resistance to reciprocity could become a major stumbling block for future dismantlement (38).

DOE has stated that enough HEU exists either in its nonweapons inventory or in warheads scheduled for retirement to meet all U.S. projected military needs for decades. DOE is currently developing plans to reconfigure the Nuclear Weapons Complex to meet these future needs (68,71). DOE's Uranium Task Force is

[24] As noted earlier, DOE has stated that some quantities of HEU may be declared surplus sometime in the future.

charged with planning for the future of its uranium operations. The task force concluded that none of DOE's weapons HEU is in excess or should be considered surplus (26), and recommended that U.S. HEU be stored for now. This would represent a stockpile for future weapons or other programs and thus delay as long as possible the need to produce more HEU for defense purposes (62). Since this recommendation, additional unilateral U.S. and Russian warhead cutbacks and Russian initiatives to sell HEU may have increased the possibility that some U.S. HEU will eventually be declared surplus and converted, although no such decision has been made (62).

Clearly, storage of weapons-derived HEU must be anticipated. Presently DOE is planning to store all of its HEU indefinitely at the Y-12 Plant (26) and is not actively considering a decision beyond such storage. Because of the prospects for U.S. purchase of converted Russian HEU (see chapter 6), all HEU issues have been discussed in that context. Not much attention has been given by the Federal Government to possible commercial uses of U.S. HEU.

DOE has recently extended the work of the Uranium Task Force in the form of an internal management plan. DOE has stated that the goal of the plan is to manage the Department's uranium resources in a manner that extends the availability of uranium to meet user needs without new production and with minimal budget outlays, while meeting new environmental, health, and safety objectives. This plan is classified, and there are no plans to produce an unclassified version (26). The plan projects uranium needs through 2005 and sets requirements for facilities in a reconfigured Weapons Complex. The uranium needs considered include national defense; fuel for tritium production reactors, naval nuclear propulsion, and space nuclear programs; research and development programs; and unspecified "commercial needs." The plan includes a model that takes into consideration these various needs and calculates the "crossover" date, the time when the need for building new production facilities could arise. The Uranium Task Force has also modeled the forms and amounts of uranium (accounting for all DOE's uranium) that will be present after reconfiguration (26).

ROBERT DEL TREDICI

DOE's Y-12 Plant, Oak Ridge, Tennessee, where highly enriched uranium from warheads is now stored.

Processing and Storage at Y-12

HEU taken from retired nuclear warheads is now stored at DOE's Y-12 facility at Oak Ridge, Tennessee. Y-12 is a large multipurpose facility with several different missions in both materials and weapons production, and a long history of working with uranium (44). In the past, the HEU components from warheads were removed and stored in special compartments at Y-12 (29).

Because Y-12 was built piecemeal, materials do not flow efficiently from place to place. The buildings are old, and there is a vast amount of waste on-site. The facility is also much larger than present or future levels of production require (40). Uranium operations at Y-12 involve many industrial processes, including casting, smelting, machining, and recycling, as well as different uranium forms (buttons, solutions, chips). Some HEU from weapons disassembled at the Pantex Plant is also processed at Y-12 (37).

DOE and the Y-12 contractors are currently reorganizing and redefining its mission—from weapons production to weapons dismantlement—

as DOE downsizes the Nuclear Weapons Complex. To improve the efficiency and cost-effectiveness of their operations, for example, Y-12 management recently reduced the number of operating uranium casting facilities from 12 to 6. Among the functions delineated in the new mission are: 1) disassembling nuclear weapons components; 2) storing and managing warhead materials such as lithium and highly enriched uranium; 3) transferring technology to the private sector (74); 4) evaluating and testing particular weapons system components; and 5) manufacturing components for other government organizations, such as the Navy's Seawolf submarine program (37).

CURRENT STORAGE ACTIVITIES

For security purposes, the area comprising the Y-12 Plant has been divided into three major zones: two low-security zones and a highly secured one. The high-security zone or "exclusion area" contains HEU processing and manufacturing facilities. This area also includes several facilities used for storing HEU and some radioactive waste generated by processing activities there (1,78).

The HEU stored at the exclusion area comes from a multitude of sources, including government and private institutions and universities. The largest volumes, however, originate from weapons disassembly operations. Upon arrival, the HEU-containing parts are inspected and temporarily stored ("staged") until the proper facilities and equipment become available to remove HEU from the containers or assemblies and prepare it for long-term storage.

When a decision is made to store HEU separated from weapons, the material is prepared for storage by recasting the metal in a specialized cylinder, placing it in a sealed container, and storing it in one of the seven operational concrete vault facilities in the high-security zone. If the HEU is part of the national strategic reserve, the container is stored in a location different from that used for nonstrategic HEU. HEU is generally stored in concrete vaults commonly known as tube vaults. Tube vaults consist of cylinders embedded in a concrete structure in a configuration that prevents any criticality accident. A typical tube vault can safely accommodate up to 40 metric tons of HEU, and its design life is estimated to be nearly 100 years (13).

In addition to HEU, Y-12 handles more than 80 other weapons materials and chemicals contained in weapons assemblies. Although HEU and certain other materials such as lithium and tungsten alloys are recycled and stored at the plant, most of the remaining inventory (e.g., aluminum, rubber, nylon, beryllium) is declassified and demilitarized before being made available to commercial facilities for recycling, treatment, or disposal. Considerable reduction in the amount of materials shipped for treatment and disposal has been achieved in the last 5 years (13).

Efforts to Address Weapons Dismantlement and Possible Impacts

Current plans call for storing HEU and other essential weapons materials returned from Pantex at Y-12's specialized storage facilities. Although the rate of "returns" has doubled since 1985, no HEU storage capacity limitations are anticipated by DOE for the foreseeable future. Since Y-12 receives only part of the total materials generated by weapons disassembly at Pantex, and since most weapons production facilities have considerably reduced their operations, plant officials claim that increases in weapons dismantlement activities will not constitute an operational or storage burden (13). Y-12 officials project current levels of personnel and expertise to be adequate for addressing future storage and processing needs for HEU from dismantled weapons.

To ensure proper management of dismantled materials, Y-12 officials have developed a computer model that estimates and projects work force needs, staging space requirements, processing and equipment demands, and long-term storage availability. Documentation detailing the handling and processing steps to be followed for each particular material returned from weapons disassembly has also been developed (13). In addition, safety analyses have been conducted at facilities where dismantlement activities take place, as well as where HEU is stored. Plant personnel are reviewing current processes and operations to determine whether additional adjustments must be made to successfully address any future dismantlement-related activities at Y-12 (8).

One possible result of expanding the storage of highly enriched uranium from dismantled weapons at Y-12 is an increase in radiation exposure during inventory assessment. Exposure levels are currently reported to be very low, particularly because of the limited ongoing processing and handling of HEU at the plant. With an increase in uranium processing and handling, exposures are expected to rise but—according to a Y-12 official—not to levels that will pose any risk to plant personnel or the general public (8).

No comprehensive analysis is available publicly that evaluates the capability of Y-12 to continue to accept and store HEU from dismantled weapons, particularly since the total quantity of U.S. HEU is classified. Plant officials do not expect Y-12 to run out of storage space for HEU. However, if such a situation developed, DOE claims that additional space could be obtained by using any of the recently closed buildings certified for HEU work. Storage space could also be made available at other facilities, but additional capital investments may be required.

Prior to a decision to use any additional existing Y-12 buildings as storage facilities for HEU, DOE will have to evaluate them in terms of safety, security, nuclear criticality, and environmental compliance. Because previous work at these facilities also involved uranium, the level of analysis required may not be extensive. Oversight by State agencies and the Defense Nuclear Facilities Safety Board may also be necessary (37). Public involvement should be incorporated in this process.

To avoid the costs associated with expanding the number of HEU storage facilities, officials at Y-12 have examined more efficient methods of storage. A new—as yet unnamed—storage system was reported to have been developed in December 1992 (13). Little public information exists, but according to Y-12 officials, the new system not only allows the storage of large amounts of HEU at subcritical conditions but is expected to triple the usable space in existing vaults.

Management and handling of HEU can lead to criticality concerns. The availability of criticality safety experts at Y-12 is limited. With the expected increase in uranium storage, efforts are being carried out to support training programs at the University of Tennessee for future staff. Several nonengineering personnel highly knowledgeable about Y-12 facilities have also been trained to become criticality safety experts. Another preventive measure being undertaken to minimize the potential for criticality safety accidents involves reducing the number of places in which HEU is handled (8).

Y-12 is one of the largest handlers of HEU in the world, and this experience could be a factor in considering a future de-enrichment and storage site should Russian weapons materials be purchased by the United States in the form of HEU.[25] Although HEU de-enrichment technologies have been employed at Y-12 for some time, its processing capacity is limited; consequently, scaling-up will be needed to handle adequately the much larger volumes of Russian HEU. The costs that may be incurred in expanding de-enrichment technologies have not been studied. In terms of storage, Y-12 officials claim to have sufficient storage space to accommodate Russian HEU, particularly in metallic form (13). If a

---

[25] As discussed in chapter 6, the current agreement to purchase Russian HEU calls for its conversion to LEU in Russia but does not preclude the possibility of future HEU shipments.

decision is made to store or process Russian HEU at Y-12, a number of technical challenges (such as the possibility of accommodating Russian monitoring) will have to be considered. It does not appear that any serious analysis has been done on this issue to date (8).

If a new storage facility is developed for plutonium, as discussed earlier in this chapter, it would be beneficial to consider HEU storage needs and criteria at the same time. Separate HEU and plutonium storage facilities may be warranted but only if the added cost and difficulty can be justified.

CHAPTER 4 REFERENCES

1. Allen, Kristen, Tennessee Department of Health and Environment, "Hazardous Waste Inspection Report," report submitted to R.J. Spence, Defense Programs, U.S. Department of Energy Oak Ridge Operations, June 22, 1989.
2. Arthur, E.D., Los Alamos National Laboratory, "Accelerator-Driven Intense Neutron Source Technology: APT and ATW," a collection of viewgraphs, Feb. 13, 1992.
3. Berkhout, F., A. Diakov, H. Feiveson, H. Hunt, E. Lyman, M. Miller, and F. von Hipple, "Disposition of Separated Plutonium," *Science and Global Security*, vol. 3, 1993, pp. 161-214.
4. Berube, Raymond, Deputy Assistant Secretary for Environment, U.S. Department of Energy, information provided to the Office of Technology Assessment at a briefing on the National Environmental Policy Act, Dec. 8, 1992.
5. Bloomster, C.H., P.L. Hendrickson, M.H. Killenger, and B.J. Jonas, "Options and Regulatory Issues Related to Disposition of Fissile Materials from Arms Reduction," Report PNL-SA-18728, Batelle Pacific Northwest Laboratories, December 1990.
6. Buckner, M.R., and P.B. Parks, Westinghouse Savannah River Company, "Strategies for Denaturing the Weapons-Grade Plutonium Stockpile," Report WSRC-RP-92-1004 prepared for U.S. Department of Energy, October 1992.
7. Buckner, M.R., et al., Westinghouse Savannah River Company, "Excess Plutonium Disposition Using ALMR Technology," Report No. WSRC-RP-92-1278 prepared for the U.S. Department of Energy, February 1993.
8. Butz, Todd, Manager, Health, Safety, Environment and Accountability, Martin Marietta Energy Systems, Oak Ridge Y-12 Plant, briefing presented to Office of Technology Assessment staff during a site visit, Dec. 16, 1992.
9. Canter, Howard, Deputy Assistant Secretary for Weapons Complex Reconfiguration, U.S. Department of Energy, personal communication, Nov. 10, 1992, and Jan. 19, 1993.
10. CEGA (Combustion Engineering/ General Atomics), "The Modular HTGR for the Destruction of Surplus Weapons-Grade Plutonium," Nov. 25, 1992.
11. CEGA (Combustion Engineering/ General Atomics), "The Modular HTGR for the Destruction of Surplus Weapons-Grade Plutonium," Mar. 15, 1993.
12. Chan, Lisa, Acting Director, Division of Engineering, Weapons Complex Reconfiguration Office, Defense Programs, U.S. Department of Energy, information provided at a meeting with Office of Technology Assessment staff, Mar. 11, 1993.
13. Chilcoat, Tony, U.S. Department of Energy, Y-12 Plant Weapons Return Program, briefing presented to Office Technology Assessment staff during a site visit, Dec. 16, 1992.
14. Claytor, Richard, Assistant Secretary for Defense Programs, U.S. Department of Energy, testimony before the Senate Committee on Armed Services, Aug. 4, 1992.
15. Clements, Tom, Southeast Nuclear Campaigner, Greenpeace, written comments, June 1993.
16. Cochran, Thomas B., Senior Staff Scientist, Natural Resources Defense Council, personal communication to Peter Johnson, Office of Technology Assessment, Sept. 2, 1992.
17. Cochran, Thomas B., Senior Staff Scientist, Natural Resources Defense Council, written comments, Office of Technology Assessment, June 1993.
18. Croff, Alan, Associate Director, Chemical Technology Division, Oak Ridge National Laboratory, written comments, June 1993.
19. Cunningham, Paul, Program Director, Los Alamos National Laboratory, information provided at the

Office of Technology Assessment workshop on "Plutonium Storage," Feb. 10, 1993.

20. Davis, Z., M. Humphries, C. Behrens, M. Holt, and W. Donnelly, *Swords into Energy: Nuclear Weapons Materials After the Cold War,* Report 92-739 (Washington, DC: Congressional Research Service, Oct. 9, 1992).
21. *Defense Cleanup,* "No New Nukes, Bush Says," vol. 3, No. 15, July 17, 1992.
22. *Defense Cleanup,* "Panel Nudges up DOE Cleanup," vol. 3, No. 11, June 5, 1992.
23. Donnelly, W.H., and Z.S. Davis, "The Role of Congress in Future Disposal of Fissile Materials from Dismantled Nuclear Weapons," paper presented at the Annual Meeting of the American Association for the Advancement of Science, Washington, DC, Feb. 18, 1991; also see Z. Davis and W. Donnelly, "Nuclear Weapons Materials: Ending U.S.-C.I.S. Production and Disposing of Inventories," Congressional Research Service Issue Brief, updated Feb. 13, 1992.
24. Drell, S., et al., MITRE Corporation, JASON Program Office A10, "Verification of Dismantlement of Nuclear Warheads and Controls on Nuclear Materials," Report No. JSR-92-331, Jan. 12, 1993.
25. Federation of American Scientists, "Ending the Production of Fissile Materials for Weapons, Verifying the Dismantlement of Nuclear Warheads: The Technical Basis for Action," June 1991.
26. Fulner, Robin, Defense Programs, and Roger Gagne, Nuclear Energy, U.S. Department of Energy, information provided during a briefing of Office of Technology Assessment staff, Aug. 27, 1992.
27. General Atomics, "The MHTGR-NPR as a Plutonium Burner," Aug. 24, 1992.
28. Guidice, Steve, Assistant Manager for Office of Operations and Weapons, U.S. Department of Energy, information provided to Office of Technology Assessment staff at a briefing on DOE nuclear weapons dismantlement, May 20, 1992.
29. Howes, Robert, U.S. General Accounting Office, personal communication, Aug. 12, 1992.
30. Johnston, Tom, Combustion Engineering/General Atomics, written comments, June 1993.
31. Kerridge, Jeff, and Leanne Smith, Office of Facility Transition and Management, U.S. Department of Energy, briefing package presented to Office of Technology Assessment staff, Mar. 11, 1993.
32. Leventhal, Paul, Nuclear Control Institute, information provided at a meeting with Office of Technology Assessment, Mar. 3, 1992.
33. Lichtenstein, Kenneth, Division of Infectious Disease, Rose Medical Center, information provided at Office of Technology Assessment workshop on "Long-Term Storage and Disposition of Plutonium from Warheads," Aug. 26, 1992.
34. Maddux, Paul, Westinghouse Savannah River Co., personal communication, Nov. 24, 1992, Dec. 30, 1992, and May 13, 1993.
35. Magwood, William, Program Manager, Nuclear Activities, Edison Electric Institute, personal communication, Aug. 19, 1992.
36. Mah, Richard, Los Alamos National Laboratory, presentation at Congressional Research Service Nuclear Nonproliferation Workshop, March 22, 1993.
37. Marrow, Margaret, Martin Marietta Energy Systems, Oak Ridge Y-12 Plant, briefing presented to Office of Technology Assessment staff during a site visit, Dec. 16, 1992.
38. McGoldrick, Fred, U.S. Department of State, information provided at meeting with Office of Technology Assessment staff, Mar. 26, 1992.
39. McPhee, John A., *The Curve of Binding Energy* (New York: Farrar, Strauss and Giroux, 1974).
40. Medalia, J., W.C. Boesman, W.H. Donnelly, and M. Holt, *Nuclear Weapons Complex: Alternatives for Congress,* Report 92-208F (Washington, DC: Congressional Research Service, Feb. 25, 1992).
41. Moore, E.N., Manager, Nuclear Materials Planning, Westinghouse Savannah River Co., personal communication, Nov. 20, 1992.
42. Moore, E.N., Westinghouse Savannah River Co., information provided during a briefing of the National Academy of Sciences Panel on Reactor-Related Options for Weapons Plutonium, Jan. 11, 1993.
43. Napier, J.M., Senior Scientist, Technics Development Corp., personal communication, Apr. 25, 1993.

44. National Research Council, *The Nuclear Weapons Complex, Management for Health, Safety, and the Environment* (Washington, DC: National Academy Press, 1989).
45. National Research Council, Board on Radioactive Waste Management, "Interim Report of the Panel on Separations Technology and Transmutation Systems," May 1992.
46. Nicks, Jim, Executive Assistant to the Assistant Secretary for Defense Programs, U.S. Department of Energy, personal communication, Nov. 18, 1992.
47. Nikipelov, B.V., Adviser to the Russian Federation Minister of Atomic Energy, information provided during an Office of Technology Assessment briefing on "High Level Nuclear Waste Management and Disposal Programs of the U.S., the Russian Federation and the Republic of the Ukraine," Apr. 23, 1993.
48. Omberg, R.P., and C.E. Walter, "Disposition of Plutonium From Dismantled Nuclear Weapons: Fission Options and Comparisons," Lawrence Livermore National Laboratory report UCRL-ID-113055 submitted to U.S. Department of Energy, Plutonium Disposition Task Force, Fission Working Group Review Committee, Feb. 5, 1993.
49. Panofsky, W., "Destroy Weapons; Make Electricity," *Bulletin of the Atomic Scientists*, May 1992.
50. Pappe, Joseph A., Jr., Colonel, Chief of Air Force Material Command Nuclear Support Office, Kirtland Air Force Base, personal communication, Oct. 10, 1992, and May 25, 1993.
51. Rask, W., Director, Rocky Flats Plant Production Division, U.S. Department of Energy, personal communication, Nov. 17, 1992.
52. Rosen, Sol, Director of International Programs, Nuclear Energy, U.S. Department of Energy, information provided at meeting with Office of Technology Assessment staff, June 29, 1992.
53. Scheinman, L., and D.A.V. Fischer, "Managing the Coming Glut of Nuclear Weapons Materials," *Arms Control Today*, vol. 22, No. 2, March 1992, pp. 7-12.
54. Sewell, Philip G., Deputy Assistant Secretary, Office of Nuclear Energy, U.S. Department of Energy, information provided during a briefing of Office of Technology Assessment staff, Apr. 27, 1993.
55. Shea, Thomas E., "On the Application of International Atomic Energy Agency Safeguards to Plutonium and Highly Enriched Uranium from Military Inventories," *Science and Global Security*, 1992, vol. 3, 1992, pp. 223-236.
56. Siebert, Bryan, Director, Office of Classification, U.S. Department of Energy, personal communication to Emilia Govan, Office of Technology Assessment, June 25, 1993.
57. Skerrett, P.J., "Nuclear Burial at Sea," *Technology Review*, vol. 95, No. 2, February-March, 1992, pp. 22-23.
58. Sutcliffe, W.G., (ed.), Center for Technical Studies on Security, Energy, and Arms Control, *Fissile Materials from Nuclear Arms Reductions: A Question of Disposition*, Proceedings for the Annual Meeting of the American Association for the Advancement of Science, Report No. CONF-910208/CTS-31-92 (Livermore, CA: Lawrence Livermore National Laboratory, Feb. 18, 1991).
59. Taylor, James, M., Executive Director for Operations, Nuclear Regulatory Commission, Policy Issue (information), SECY-92-064, Feb. 25, 1992.
60. Taylor, J.J., and E. Rodwell, "Disposal of Fissionable Material from Dismantled Nuclear Weapons," Electric Power Research Institute, unpublished paper, May 1992.
61. U.S. Congress, General Accounting Office, "Briefing Report to the Honorable David E. Skaggs, House of Representatives—Summary of Major Problems at DOE's Rocky Flats Plant," GAO/RCED-89-53BR, October 1988.
62. U.S. Congress, General Accounting Office, "Uranium Enrichment, Unresolved Trade Issues Leave Uncertain Future for U.S. Uranium Industry," GAO/RCED-92-194, June 1992.
63. U.S. Congress, General Accounting Office, "Nuclear Materials—Removing Plutonium Residues from Rocky Flats Will Be Difficult and Costly," GAO/RCED-92-219, September 1992.
64. U.S. Congress, Office of Technology Assessment, *Complex Cleanup: The Environmental Legacy of Nuclear Weapons Production*, OTA-0-484 (Washington, DC: U.S. Government Printing Office, February 1991).
65. U.S. Congress, Office of Technology Assessment, *Long-Lived Legacy: Managing High-Level and Transuranic Waste at the DOE Nuclear Weapons*

Complex, OTA-BP-0-83 (Washington, DC: U.S. Government Printing Office, May 1991).

66. U.S. Congress, Office of Technology Assessment, workshop on "Long-Term Storage and Disposition of Plutonium from Warheads," Aug. 26, 1992.
67. U.S. Congress, Office of Technology Assessment, workshop on "Plutonium Storage," Feb. 10, 1993.
68. U.S. Congress, Office of Technology Assessment, "Subseabed Disposal of High-Level Radioactive Waste," staff paper, May 1986.
69. U.S. Congress, Office of Technology Assessment, *Proliferation of Weapons of Mass Destruction: Assessing the Risks* (Washington, DC: U.S. Government Printing Office, forthcoming).
70. U.S. Department of Defense, Department of the Air Force, Air Force Material Command, Nuclear Support Office, "System Safety Program Plan: Kirtland Underground Munitions Storage Complex," October 1992.
71. U.S. Department of Energy, Albuquerque Operations, Amarillo Area Office, "Environmental Assessment for Interim Storage of Plutonium Components at Pantex," predecisional document, DOE/EA-0812, December 1992.
72. U.S. Department of Energy, *Nuclear Weapons Complex Reconfiguration Study*, DOE/DP-0083 (Washington, DC: U.S. Department of Energy, January 1991).
73. U.S. Department of Energy, Office of Defense Programs, *Implementation Plan: Nuclear Weapons Complex Reconfiguration Programmatic Environmental Impact Statement*, DOE/EIS-0161IP (Washington, DC: U.S. Department of Energy, February 1992).
74. U.S. Department of Energy, Office of the Press Secretary, "Departments of Energy, Commerce to Aid Machine Tool Industry," press release, Mar. 5, 1993.
75. von Hippel, F., "Bombs Away," comment in *Discover*, April 1992, *Defense Cleanup,* vol. 3, No. 2, Jan. 31, 1992, p. 1.
76. von Hippel, F., "Control and Disposition of Nuclear Weapons Materials," talk at Fuel Cycle 92, Mar. 11, 1992.
77. von Hippel, F., D.H. Albright, and B.G. Levi, "Quantities of Fissile Materials in the U.S. and Soviet Nuclear Weapons Arsenals," Center for Energy and Environmental Studies, July 1986.
78. Waddle, Susan, U.S. Department of Energy, Y-12 Plant Site Manager, briefing presented to Office of Technology Assessment staff during a site visit, Dec. 16, 1992.
79. Westinghouse Savannah River Company, "Vitrification of Excess Plutonium," WSRC-RP-93-755, Predecisional Draft, Aiken, SC, May 1993.
80. Willet, Lou, Deputy Director, Office of Weapons Materials and Programs, Defense Programs, U.S. Department of Energy, information provided during briefing of Office of Technology Assessment staff, Mar. 11, 1993.

Public Concerns and Information Access 5

Dismantling warheads and managing the special nuclear materials from them pose great challenges to the United States and to the Government institutions that will be charged with a variety of complex tasks. A prerequisite for effective action is a recognition that conducting this work in the post-Cold War context is fundamentally different in terms of mission from warhead production during the arms race. Although some physical tasks may remain the same, they are now part of a new mission that involves permanent reduction of the nuclear weapons arsenal. A crucial component of this mission is the need to provide responsible stewardship of nuclear materials from dismantled warheads and to develop solutions for ultimate disposition of these materials in a safe manner that protects the environment and human health.

Although the world situation has changed dramatically, the institutional context in which nuclear weapons policies are developed remains largely as it was throughout the Cold War. Policy decisions are still being made within the old legislative, administrative, and cultural framework, and the United States has failed to develop a national consensus that could lead to a focused, new policy and provide the basis for a clear, new mission. The present institutional framework may not be appropriate to fulfill competently the responsibilities involved in the new mission.

Discussion and debate about the role of nuclear weapons in national security continue to be conducted in two largely separate and distinct arenas. One is dominated by members of the national security establishment—particularly those with access to classified information—who develop, evaluate, or implement national

Point

"What would be gained from a public debate on the issue on the specifics of how the DOD determines which capabilities it will maintain in light of international trends and treaty obligations?"

Pentagon reviewer of OTA report

Counterpoint

"Site by site, citizens will have to join together to force the DOE to change by saying that they will not entrust the next 40 years to the same regulatory and bureaucratic structure that created the last 40 years."

Local citizen group reviewer of OTA report

policy. The other is populated by outsiders, supporters, or critics of national policy (including the media and academia), who have opinions but very limited access to relevant information.

As described in chapter 2, decisions about the size and makeup of the nuclear weapons stockpile are made by a Nuclear Weapons Council composed of officials from the Departments of Energy (DOE) and Defense (DOD). Each year the Council prepares a stockpile memorandum that, when approved by the President, determines the amount and status of warheads for the next 6 years. Like most other matters concerning nuclear weapons and nuclear materials, these analyses and deliberations are conducted in secret and the results are disclosed only to those with appropriate security clearance. The total number of weapons in the current U.S. nuclear stockpile is classified, as are the numbers of warheads in either the active stockpile or the inactive reserve (10).

On the operational level, most decisions about U.S. warhead dismantlement plans and logistics are made jointly by DOD and DOE. Decisions about specific activities are made by each agency individually. Primary responsibility for producing nuclear warheads rests with DOE; warhead production programs are thus carried out apart from the military agency that is the customer for nuclear weapons—DOD. After they are assembled, DOD takes custody of warheads from DOE and retains them until they are retired; custody is then transferred back to DOE for dismantlement and the disposition of nuclear materials.

Some public interest groups are concerned that the decisionmaking process regarding nuclear weapons policies does not permit adequate and informed public debate on important national issues. The most fundamental matters involve the redefinition of international security in light of the changed world situation, and the role of nuclear weapons in preserving U.S. national security. Specific policy matters that are not being addressed publicly include: 1) the number of nondeployed nuclear warheads that the United States and Russia intend to retain as "backups" for the deployed warheads allowable under current agreements, 2) the number of warheads planned for retirement and dismantlement over the next two decades, 3) the amount of nuclear material from dismantled warheads, and 4) the ultimate disposition path for nuclear material that is deemed unnecessary for defense purposes.

Another concern often stated is that, because DOE has exclusive jurisdiction over all matters relating to special nuclear materials,[1] it is not subject to outside scrutiny in these matters (3). This arrangement distinguishes weapons-related activities from atomic energy activities in the civilian or commercial sector, which are overseen by the Nuclear Regulatory Commission and other appropriate agencies (3). Within DOE, nuclear weapons responsibilities are housed in the Office of the Assistant Secretary of Defense Programs (DP). The office, although part of a civilian agency, has a predominantly military mission. By law, responsibility for the production, storage, and accounting of special nuclear materials is carried out by a deputy assistant secretary for military applications from the military ranks (11). Other offices in DOE support defense program activities or are customers for the nuclear materials produced under its auspices,[2] and some national laboratories play an important role in all aspects of nuclear weapons research and development.

[1] Special nuclear materials are defined by the Atomic Energy Act as "plutonium, uranium enriched in the isotope 233 or in the isotope 235, and any other material which the Commission . . . determines to be special nuclear material . . . but does not include source material" (42 U.S.C. 2014). The International Atomic Energy Agency defines "special fissionable material" as "plutonium-239; uranium-233; uranium enriched in the isotopes 235 or 233; any material containing one or more of the foregoing; and such other fissionable material as the Board of Governors shall from time to time determine" (17).

[2] For example, the Office of Nuclear Energy (NE) often acts as a "customer" for materials produced by Defense Programs. NE conducts research and other activities with the materials.

PUBLIC CONCERNS

Public interest groups, concerned about weapons dismantlement and the disposition of plutonium and highly enriched uranium (HEU) from dismantled warheads, want to participate in national decisions on this matter. They recognize that one of the key issues is whether these materials are considered a resource to exploit, as in power reactors, or a waste requiring disposal. Regarding plutonium, many believe that the costs and waste streams associated with its possible use in the U.S. nuclear energy industry would be prohibitive.

Regardless of the final outcome of this debate, public interest groups realize that an interim storage period will be required, and they are concerned about potential health and environmental threats from storage of these materials, especially plutonium. They are also concerned about possible international proliferation problems. Domestically, their concerns include: 1) the adequacy of plutonium interim storage containers to reasonably ensure health and human safety, 2) the insulation of interim storage sites from natural disasters or other scenarios that may endanger the surrounding area, and 3) the need for independent oversight of storage facilities and operations. Until such questions are answered, the public will remain unconvinced of the safety of DOE's interim storage plans.

A long and deeply ingrained distrust of DOE exists among public interest groups. This distrust stems primarily from the sense of frustration the public has from past experience in dealing with DOE on environmental issues. The concerns range from the responsiveness of DOE in meeting requests from the public for information, to a perception of disregard for public safety and environmental integrity in deference to production goals. These groups believe that, to some extent, DOE also distrusts the public, and that this is most likely the result of an institutional culture that has put a premium on security and secrecy. DOE, they feel, is simply not accustomed to operating and making policy decisions in an open and public manner, which has had a negative impact on its relationship with some sectors of the U.S. public. The Office of Technology Assessment (OTA) informally surveyed public interest groups to determine recent experiences in obtaining information about DOE facility operations. A summary of results is shown in box 5-A.

Regardless of the sources or extent of the mutual distrust, it has manifested itself as a strong concern by the public over whether DOE—the agency responsible for creating the Nation's warheads—is the best institution for dismantling them as well. Most public interest groups near DOE sites affected by dismantlement see a need for greater public involvement in decisions affecting dismantlement operations. They strongly support the concept of including the public earlier and more actively in relevant decisionmaking processes. Despite recent efforts by DOE to respond to these views, public interest groups feel that effective and meaningful public involvement is not being achieved. In fact, the consensus is that negligible progress has been made toward a true engagement of the public and that DOE continues to ignore public concerns.

A recent performance evaluation of the operations and maintenance contractor at Pantex confirms this appraisal, noting a significant lack of management attention to public affairs. During the evaluation period, the contractor, Mason & Hanger-Silas Mason Co., Inc. (M&H), was supposed to demonstrate improvement in a proactive, public affairs and community relations program, including accurate and early identification of issues. M&H did not meet this criterion and expended little effort in the area, apparently giving it low priority. The evaluation (35) noted that the:

> Pantex Plant's role in downsizing the nuclear weapons stockpile has placed it in the world spotlight. Media, public and activist group interest continues to grow. This facility is slowly losing ground in terms of public acceptance and

the trend will continue at an ever increasing rate without a concerted effort to develop and implement a comprehensive public affairs program that continually seeks to present an accurate picture of Pantex Plant functions to the public.

Incidents highlighting credibility problems at Pantex are described in box 5-B.

Site-Specific Concerns

Activities at DOE Weapons Complex sites affect surrounding communities in a variety of ways. Historically, they have provided employment for large segments of the population in adjacent towns and cities. However, many believe that local economic benefits have come at the expense of a negative environmental legacy. Approximately 45 years of nuclear weapons production, with its associated materials fabrication and processing plants, reactors, and nonnuclear component needs, have caused widespread contamination that has led to considerable public concern about health and the environment.

Because of these integral links to communities that both support and depend on the Weapons

Box 5-A—Public Participation and Access to Information

Weapons dismantlement activities will affect a variety of people who live in areas hosting DOE weapons facilities. As a result of the presence of these facilities, and the potential hazards associated with them, many local citizen groups have formed that seek to learn about, monitor, and participate in their planning and operations. Although specific concerns may vary depending on the site and its associated activity, in general, these groups seek to participate in the decisionmaking process to ensure that adequate precautions are taken to protect public health, worker safety, and the integrity of the environment.

The efforts of these citizen groups have been impaired, however, by the information classification system, as well as by an ineffective public communications and inquiry structure and the lack of processes to meaningfully consider and respond to public concerns in DOE's decisionmaking. OTA informally surveyed a limited number of citizen groups about their experiences with information requests to DOE regarding environmental, safety, and health issues. The following responses characterize common problems.

- A fall 1992 request from a Hanford, Washington group about leach and leak testing, hydrogen generation, and sampling in a Hanford site grout program was responded to in a timely manner; however, "much information was withheld under 'predecisional' status. This has occurred several times with my Freedom of Information Act (FOIA) requests. DOE appears to circumvent the law by keeping documents in draft form for extended, unwarranted periods of time" (18).
- At the Savannah River Site, a group documented many requests for information related to plutonium operations at the facility. Some inquiries, initiated in 1990, are still pending final responses. The respondent indicated that "greater headquarters involvement in and control over DOE activities has created a bottleneck which often significantly delays the release of information" and "DOE relies heavily on FOIA exemptions, as well as classified and unclassified, controlled nuclear information labels. These mechanisms make it difficult to identify documents related to public concerns and create substantial delays in the release of information" (6).
- A group located near the Idaho National Engineering Laboratory (INEL) requested a large amount of information on radioactive and chemical releases from operations and accidents, as well as worker radiation exposure records. After submitting a FOIA request, the DOE review concluded that the group be charged $1,227,900 for fulfillment of the request. According to the group, "this figure has more to do with the line item budget requirements than with the actual costs involved." The perception is that "DOE has used every tactic imaginable to frustrate the release of information on INEL" (4).

- Differences have been reported in the timeliness and openness of DOE, depending on the office involved. For example, a Hanford group stated that "the staff of the Richland Field Office are more responsive than those at Headquarters." Also, differences among *divisions* of DOE were reported: "It always takes longer with DP [defense programs] matters. EH and EM [Environment, Safety, and Health and Environmental Restoration and Waste Management, respectively] officials are almost always easier to deal with." The overall assessment was that "DOE does not assign sufficient resources to classification review" (31).
- Even when responses are timely, the information provided is often not directly related to the request, or is insufficient. In response to a FOIA request by a national group concerning a DOE work order, DOE sent ". . . a nearly blank page, except for one sentence at the bottom which read, in its entirety, '6. *Analysis of Public Issues* will include an analysis of public perception of health and safety risks, and perception of potential economic risks (e.g., contamination of local crops). The rest of the work order was never declassified, even in part." After pursuing the issue for more than 8 months, "DOE has provided . . . nothing" (27).

These anecdotes illustrate the feelings and perceptions of citizen groups about information accessibility at DOE. Though by no means a comprehensive analysis or sample, they illustrate problems. Obviously, not every request for information is denied, ignored, or delayed. However, the security environment in which DOE is legally bound to operate, and the closed nature of many operations, too often hamper legitimate efforts of the public to educate themselves about operations at DOE facilities. Whether intended or not, the effect is that public confidence is eroded. The new Administration has told citizen groups it will give a high priority to solving these problems.

SOURCE: Office of Technology Assessment, 1993.

Complex, many public organizations have been formed that seek to address the environmental and health issues associated with these sites. Some groups are now also addressing the environmental impacts of new warhead dismantlement and materials management activities.

Virtually all public interest groups concerned with nuclear weapons issues, and dismantlement specifically, share common concerns about DOE's competence, the public's access to relevant information, and effective citizen involvement in national decisionmaking. Although usually organized in response to concerns about a local site, these groups also recognize and engage the national scope and international implications of Weapons Complex issues. However, in addition to common concerns about DOE's operations, public interest groups have problems specific to the key dismantlement sites: Pantex, the Y-12 Plant, and the Savannah River Site.

PANTEX

Several public groups are involved with, and concerned about, activities at Pantex. PANAL, Panhandle Area Neighbors and Landowners, is a group of Pantex area residents who organized when Pantex's mission changed to one of disassembly and storage. The Peace Farm—an area adjacent to the facility—serves as a community for arms control and disarmament advocates. Serious Texans Against Nuclear Dumping (STAND) was initially formed in opposition to siting a nuclear waste repository in Texas (29). Pantex activities, especially relative to dismantlement, are now the main concern of STAND. The Texas Nuclear Waste Task Force is a coalition of groups with various agendas that was formed to resist the siting of a high-level nuclear waste repository in Texas. All these groups have joined to form STAR, the State of Texas Alliance for Resources, to address concerns that they share about Pantex (14).

The group that has made Pantex dismantlement and storage programs the center of its activities is

BEVERLY GATTIS

Local citizens living near the Pantex Plant set up an information booth during DOE hearings on Environmental Impact Assessment.

STAND. STAND's current efforts focus on stopping the open burning of chemical explosives and stopping Pantex from becoming the long-term storage site for plutonium pits. STAND and others believe that open burning of explosives may represent a public health threat. Of particular concern is fluoride deposition in soil in the immediate area, which serves cattle grazing and other agricultural uses. The future plutonium pit storage issue is particularly worrisome to STAND because Pantex is already storing large numbers of pits. DOE is now planning to increase pit storage capacity in Pantex bunkers (see appendix A). Some in the local community are concerned that this will increase radioactivity within the storage area and, thus, increase worker safety risks. STAND, as well as the Texas Attorney General, is also concerned that no plans are being made to designate an alternative site for pit storage, which indicates that a decision to use Pantex for a more extended time may occur by default. Other concerns of the public interest groups around Pantex include the impact of increased waste streams from heightened disassembly activity, the potential contamination of a large groundwater aquifer in the region, and worker safety.

Y-12 AT OAK RIDGE

The primary public interest group associated with the Y-12 site is the Oak Ridge Environmental Peace Alliance (OREPA). The Alliance acts as a public educator about activities at the Oak Ridge Reservation as well as an advocate for changes in many aspects of DOE's operations there (22). Regarding dismantlement programs, the Alliance is concerned mainly with the new waste streams occurring at Oak Ridge, particularly at the Y-12 Plant. Issues of storage capacity and safety, proper treatment of components classified as waste, and long-term disposition of uranium from warheads are top priorities. The group is urging DOE to prepare an Environmental Impact Statement (EIS) for dismantlement activities at Y-12, or at least have a section of DOE's Programmatic EIS for reconfiguration devoted to these issues. The waste stream concern is particularly acute since the site is already listed on the Superfund National Priorities List for cleanup.

OREPA also sees its role as that of a "conscience" for DOE. As dismantlement progresses, the group will attempt to verify what information DOE releases, assess its accuracy, and prod DOE into maintaining a dialogue with the public. This latter goal derives from distrust of DOE among local residents, in light of recent revelations about mercury contamination at the Oak Ridge site. In 1983, it was reported that more than 2 million pounds of mercury had been released to the environment during Oak Ridge operations in the 1960s (22). This incident is one of many that have generated widespread distrust of DOE by the public and have led to the conviction of public interest groups that they must act as oversight bodies.

Another OREPA concern is DOE's lack of a formal plan for the role of Y-12 in its dismantlement program. The group is concerned that Y-12 may become a de facto storage site for hundreds of tons of highly enriched uranium. OREPA wants to avoid this and seeks to engage DOE, the State of Tennessee, and other relevant agencies in

Box 5-B—Credibility Problems at Pantex

In July 1992, a Texas Water Commission official accidentally discovered, when determining whether classified hazardous waste had been stored in a bunker marked for pit storage, that the plutonium pits were stored in a different configuration than Pantex has consistently represented to the State (23). Instead of drums in rows arranged along the sides of the bunker, the pits had also been placed in the middle. The specified 4-foot aisle access space, emphasized by Pantex officials as a protective measure for workers during monitoring and inventorying operations, was lacking, and the Pantex official entering the bunker had to walk sideways. The Pantex official claimed that the represented configuration was typical and the particular bunker was experimental. Three workers entering the bunker did not put on any protective clothing, even though they seemed to know it was required. The observed configuration did not appear optimum for worker safety. Moreover, Pantex officials appeared uncertain about the actual number of pits stored in each bunker (23). The varied explanations given by Pantex for the aberrant configuration do not inspire confidence and appear to be after-the-fact justifications (13).

In April 1992, radioactivity was detected at the high-explosive burning grounds. DOE apparently conducted an internal investigation but failed to notify the State. The State found out indirectly, several months later, from a citizen who had heard it secondhand from a Pantex worker. Since the State was not informed of the incident when it occurred, it was not given the opportunity to participate in the subsequent investigation. The State-DOE relationship is not enhanced when an incident occurs and State and local officials find out about it months later from concerned citizens or the local newspaper (13).

SOURCE: Office of Technology Assessment, 1993.

dialogue with the public in an open decision- and policymaking process.

SAVANNAH RIVER SITE

The Energy Research Foundation (ERF), located in Columbia, South Carolina, is the lead public interest group associated with the Savannah River Site. ERF also addresses nationwide Weapons Complex issues (6).

Public concerns about dismantlement programs at Savannah River include potential plutonium storage and the storage of tritium canisters. Although current plans do not call for pit storage at the site, the option has been discussed (plutonium from past production operations is currently stored there). Tritium canisters, which are fabricated at Savannah River, are being returned and stored there as warheads are dismantled. ERF has concerns about ongoing tritium recycling activities that have, in the past, released tritium directly into the environment and generated tritium-contaminated waste. Finally, the group wishes to participate in decisions that may flow from current study proposals to build reactors or other facilities at the Savannah River Site for plutonium disposition.

■ National Public Interest Group Concerns

In addition to the site-specific organizations described above, a range of national interest groups also are concerned with dismantlement issues. Two of the groups, the Military Production Network and the Plutonium Challenge, are coalitions of national environmental organizations and locally based citizen groups. Another national group concerned with dismantlement issues is the Natural Resources Defense Council (NRDC).

The Military Production Network (MPN), with an office and full-time representative in Washington, DC, describes itself as an alliance of 41 grassroots and national organizations that addresses issues of nuclear weapons production and waste cleanup. Given the change in mission of many of DOE's weapons sites, MPN has also turned its attention to issues associated with dismantlement. STAND and OREPA are members of MPN, and ERF is listed as a "friend" of MPN (19).

MPN seeks to influence national policy by working with Congress, the Administration, and specifically, regulatory agencies and DOE on a range of issues. MPN has issued a formal position on dismantlement, which calls for DOE to make public all plans and information regarding storage of plutonium and highly enriched uranium from warhead dismantlement, as well as the use of plutonium as reactor fuel or decisions on its final disposition. MPN opposes the disposal of transuranic and mixed transuranic wastes at the Waste Isolation Pilot Plant, and calls for more public involvement and environmental sensitivity in the DOE decisionmaking process.

The Plutonium Challenge, organized in 1986, began as a coalition of arms control and environmental groups that supported a ban on the production of weapons-grade plutonium. Recently, it has widened its agenda and addresses several issues concerning the Weapons Complex and nuclear arms. The coalition meets each week to develop legislative strategies aimed at congressional action. The Plutonium Challenge intentionally limits its focus to a few key issues, and dismantlement is currently not a priority. Although the coalition has not issued or announced a formal stance on dismantlement, it monitors events and considers the impacts of weapons dismantlement on its overall agenda (9).

The Natural Resources Defense Council, founded in 1970, is a public interest group composed of a staff of attorneys, scientists, engineers, and public policy specialists with expertise in environmental, energy and resources, economic, and proliferation and disarmament issues. NRDC works on a wide variety of issues in these fields, and has initiated efforts to address warhead dismantlement and nuclear materials disposition as well (5).

NRDC has long focused on nuclear weapons policy and planning, and is considered an authority on the history and processes related to nuclear weapons production. Its *Nuclear Weapons Databooks* are widely recognized as the authoritative information source for nuclear weapons issues. Recently, NRDC, in cooperation with the Federation of American Scientists, has hosted a series of international workshops on dismantlement. These meetings highlighted a number of issues that were later pursued in negotiations with former Soviet officials by U.S. Government agencies.

ROBERT DEL TREDICI

Citizen activists from the United States and Russia meet at the DOE Savannah River Site.

One of NRDC's chief interests within dismantlement policy is that a system of verification and information exchange between the United States and the former Soviet Union be developed to address the risks of fissile materials proliferation. The long-term disposition of special nuclear materials from dismantlement is also a primary concern of NRDC. Like the local groups, NRDC places a premium on the openness of, and public involvement in, the nuclear weapons dismantlement decisionmaking process.

INFORMATION ACCESS

To control the dissemination of information that could threaten national security, certain restrictions on access to "atomic energy" information have been established. Certain information and data regarding nuclear weapons activities must be protected to prevent the proliferation of nuclear weapons, as well as terrorist threats such as the theft of special nuclear materials or weapons, the diversion of these nuclear materials, or sabotage of nuclear weapons facilities.[3] These restrictions were established in the Atomic Energy Act (AEA) (42 U.S.C. section 2011-2296 (1982 and Supp. IV 1986)) and its amendments, as well as in the security classification systems subsequently developed by DOE and DOD. Both agencies carry out their respective nuclear weapons missions under complicated systems of information classification and security. These systems are based on a comprehensive set of laws, Executive orders, and internal rules and orders.

The primary legal foundation for DOE's information classification and security system is the Atomic Energy Act of 1954, as amended (3). The act defines Federal agencies' obligations with respect to controlling information related to atomic energy defense programs. DOE also works with "National Security Information," which is regulated under an Executive order; however, atomic energy information is controlled exclusively by the act (12).

Under the AEA, a broad scope of information related to atomic weapons and processes involving special nuclear materials is categorized as "Restricted Data"[4] and deemed classified from the moment it is produced (i.e., it is "born classified") (16). This condition is unique to nuclear information under the control of DOE. Such data may be declassified only by a positive action of DOE. In the case of information that has been removed from the Restricted Data category and placed under joint control of DOE and DOD ("Formerly Restricted Data," related primarily to the military utilization of atomic energy), a decision to declassify must be made jointly by the two agencies. DOE also issues its own orders delineating procedures and guidelines for handling information classification and security.[5]

Restricted Data, once produced, remain classified indefinitely. That is, there is no expiration date beyond which such information becomes unclassified. The Atomic Energy Act does, however, mandate that classified information be continuously reviewed and declassified when conditions merit (3).

DOE has no office or formal organization that deals exclusively with declassification,[6] but it does have a process for declassification of Restricted and Formerly Restricted Data (34). The Office of Classification within the Office of Security Affairs issues biennial calls for declassification proposals from DOE programs, field and operations offices, and contractors. (Since 1990, the Office of Classification has issued two biennial calls.) The Office of Classification then reviews the proposals to determine whether the requesting office has adequately justified the action.[7] Many criteria are used to judge whether classified information may be declassified, including:

> the extent to which the information would assist in the production of special nuclear material, . . . the benefit to be realized by the U.S. program from the declassification action . . . , [and] "the cost to the U.S. program by the continued classification of the information (34).

---

[3] For a thorough description of why information requires classification protection, see reference 25.

[4] "The term 'Restricted Data' means all data concerning (1) design, manufacture, or utilization of atomic weapons; (2) the production of special nuclear material; or (3) the use of special nuclear material in the production of energy, but shall not include data declassified or removed from the Restricted Data category pursuant to section 2162 of this title" (42 U.S.C. section 2014(y)).

[5] DOE Order Series 5600 provides guidelines for DOE personnel working with classified or controlled information.

[6] The Office of Classification within the Office of Security Affairs, which is part of the Office of Intelligence and National Security, is the organizational element with responsibility for classification, declassification, and Unclassified Controlled Nuclear Information.

[7] The biennial call declassification process is the normal procedure for declassification actions. However, departmental elements may submit requests for declassification reviews at any time.

A schedule for accomplishment of the review process is also set by the Office of Classification.

The recommended declassification actions are then distributed to all offices and organizations involved and, if appropriate, to the Department of State and the Arms Control and Disarmament Agency for input on proliferation concerns. Comments from these reviewers, as well as the Office of Classification, are then forwarded to a Technical Evaluation Panel, which reviews the proposals and recommends either for or against the requested action. The panel consists of three nuclear weapons experts, one from each of the weapons laboratories: Sandia, Lawrence Livermore, and Los Alamos (15). The recommendations of the Technical Evaluation Panel are then forwarded to the Office of Security Affairs for approval or disapproval. Approved declassification actions are implemented through internal DOE bulletins or revisions to classification guidelines (34).

Although the Office of Classification is the organization with authority over classification, declassification, and unclassified but controlled information within DOE, it is not the only departmental element with influence on classification *policy*. Most of the classified information or material within DOE is ''owned'' by Defense Programs, which exercises a degree of control over all declassification decisions (37).

In addition to classified information, DOE has special procedures for information that is not classified but is judged to be ''sensitive.'' Several categories of unclassified, yet limited-use, information exist. The most important category with regard to weapons dismantlement is ''Unclassified, Controlled Nuclear Information'' (UCNI), which is related to nuclear activities and was promulgated in regulations required by the Atomic Energy Act. The AEA prohibits unauthorized dissemination of UCNI, and regulations (10 CFR 1017) specify the legal conditions for dealing with such information.[8] Generally, UCNI is not available to the public unless one requests, and justifies, ''special access'' under the provisions of the law.

The need for the UCNI category has been questioned by some (1,20,40). Given the sweeping nature of AEA's information classification authority, as well as other governmental provisions for information security, UCNI has been criticized as redundant and unnecessary for the adequate protection of information on atomic energy defense programs. A key difference between UCNI and most other classified DOE information is that data are categorized as UCNI after a judgment is made by DOE, rather than as a presumptive condition of the material (as with Restricted Data).

In one case, a document was categorized UCNI and, after heightened interest and inquiries from citizens and public interest groups, released to the public in a ''sanitized'' form.[9] Apparently, the sanitized version differed from the UCNI document very slightly, which indicated that, with minimal effort, a version could have been made publicly available at the same time the DOE version was completed. Had this occurred, DOE could have made progress in establishing positive communication and openness with the public, and improved the level of trust and credibility with the public. However, the need for continued inquiry and pressure from citizens and public groups frustrates the process and damages the image of DOE.

DOE also handles information classified under security provisions other than the AEA. The legal basis for other than Restricted Data and Formerly Restricted Data categories is Executive Order 12356, issued by President Reagan in 1982 (12). Executive Order 12356 governs the classification of ''National Security Information.'' National

[8] DOE Orders 5635.4 and 5650 describe agency procedures for identifying and protecting UCNI.

[9] The document in this case was a Safety Analysis Report conducted to analyze risks and conditions associated with increased storage of plutonium pits at the Pantex facility.

Security Information differs significantly from Restricted Data and Formerly Restricted Data in that it requires a positive action by an agency or authorized official to classify something. National Security Information is defined by Executive Order 12356 as falling into 10 categories, including:

> (1) military plans, weapons, or operations; (2) the vulnerabilities or capabilities of systems, installations, projects, or plans relating to the national security; . . . or (7) United States Government programs for safeguarding nuclear materials or facilities.

Thus, it is clear that Executive Order 12356 can also apply to DOE Weapons Complex activities and operations (but not to atomic energy information encompassed by the Atomic Energy Act).

President Clinton has recently directed a comprehensive review of the National Security Information classification system that will culminate in the preparation of a new Executive order to replace Executive Order 12356. The Review Directive states that:

> [W]e should re-evaluate our security classification and safeguarding systems . . . to ensure that they are in line with the reality of the current, rather than the past, threat potential (24).

The Directive outlines specific questions to be addressed as part of the review, including: What steps can be taken to avoid excessive classification? What steps can be taken to declassify information as quickly as possible? It is important to distinguish, however, that the Presidential Review Directive addresses National Security Information and not Restricted Data (which is regulated by statute). Thus, the status of information such as nuclear materials stockpile amounts and numbers of weapons in the stockpile or slated for retirement and dismantlement will not change as a result of the review.

Besides the legal foundation that restricts access to information, the limited infrastructure and resources devoted to information classification and declassification at DOE hamper effective and timely response to public information requests. Responses to requests for information are often slow and insufficient. Information and documents that are disseminated are too often released only after great time and effort have been expended by the requester. Even when the final action is a denial due to classification of the information, it is not uncommon for DOE to take months, or sometimes years, to respond to a request (6,18,26,27).

Undoubtedly, a large part of this problem is due to the constraints under which the Office of Classification operates with respect to financial and human resources. The Office of Classification has been level-funded since 1980, and recently experienced a significant reduction of funding in relation to its total budget. These cuts have had a particularly adverse impact on its ability to respond to Freedom of Information Act (FOIA) (Public Law 89-487) requests.[10] Although mechanisms do exist for the routine review of classified material, and its subsequent declassification if deemed necessary, they are overloaded.[11]

The statutory and practical restrictions on information access also affect the accessibility to the public of the Defense Nuclear Facilities Safety Board (DNFSB), which has the authority

---

[10] The Freedom of Information Act, which allows citizens to request access to government information not generally made available, and to challenge the withholding of information, has an exemption for "properly classified" material. Although FOIA places the burden on the "owner" agency to justify the denial of access to information or documents, all properly classified material is exempt from provisions of the act. FOIA requests are sometimes responded to by releasing an unclassified version of the restricted material, complete with blacked-out sections of text or charts.

[11] In 1992, the Office of Classification reviewed approximately 150,000 classified documents categorized as environmental, health, and safety information. The number of people engaged in and authorized for review activities is approximately 100 (37). In contrast, more than 5,000 people agency-wide have some type of classification authority (30).

to oversee nuclear weapons facility activities and advise DOE about improving the safety of those activities. Citizens have found that many documents generated by the Board are categorized by DOE as UCNI (which are available only after they have been "sanitized"). Others are difficult to get, often requiring a FOIA request. The Board also keeps a restricted database to which the public does not have access (7). In 1990, a lawsuit was filed against the Board that "challenged the Board's position that it was not an 'agency' for purposes of the Sunshine Act and the Freedom of Information Act" (36). A Circuit Court of Appeals ruled that the Board must be considered an "agency" and must therefore develop rules for complying with the acts. However, the Board was allowed to hold closed meetings on recommendations regarding safety and health at DOE defense nuclear facilities since its enabling statute includes language interpreted by the Court to allow this (36).

In addition to affecting public access to information such as numbers of nuclear weapons and materials, security limitations hinder access to information about the environmental, safety, and health aspects of activities at DOD facilities that house nuclear weapons; at DOE's Nuclear Weapons Complex; and with respect to the transportation of nuclear weapons and materials. Recent investigations suggest that the process for implementing nuclear weapons policies—which has historically given priority to national security considerations, at the cost of neglecting the protection of human health and safety, and the preservation of the environment—has not greatly improved (2,21,28,32,38).

National public interest groups and citizens near certain Weapons Complex sites are complaining that current rules restrict their access to environmental, health, and safety data that have little to do with national security. At the local level, the failure to disclose such data has contributed to a lack of trust of Federal agencies and has promoted an adversarial relationship between DOE and its contractors on the one hand, and States, community groups, or other interested parties on the other. There are concerns that it may also allow the environment, health, and safety to continue to be relegated to a low operational priority in Weapons Complex activities.

Recently, the Office of Classification has undertaken initiatives to address the issues and problems described (see box 5-C). A draft Classification Policy Study was completed that made several recommendations aimed at modifying the classification environment to adapt to new international security conditions. Also, a department-wide environmental, safety, and health initiative included a directive to review such information and to prepare future environmental, safety, and health documents with "an eye toward public release" (33).

In sum, DOE now has discretion to limit access to a broad range of information relevant to weapons dismantlement and nuclear materials management. As a result, the public's ability to acquire adequate and timely information regarding environmental, safety, and health issues related to these activities is greatly impaired. Information that citizens consider essential to discussions of safety and health is often inaccessible to interested persons outside DOE because it is classified or otherwise tightly controlled. Citizens frustrated by lack of access to information are not likely to trust the agency or support its plans and programs (39). Yet such trust and support are critical if warhead dismantlement and materials disposition programs are to gain needed public acceptance.

APPROACHES FOR INSTITUTIONAL CHANGES

If progress is to be made toward warhead dismantlement and sound materials disposition, it will be necessary to move from the present situation of scattered ideas and initiatives to a new approach of developing broadly acceptable goals and objectives on which to carry out focused solutions and attain desired results. A major

Box 5-C—Recent Initiatives Regarding Information Access

President Clinton, as well as the Department of Energy, has recognized the need for modifying the system under which government information is classified and controlled. Greater public interest and concern about the Nation's defense programs (including nuclear operations) domestically, as well as fundamental changes in the international security environment, have pointed out the need for a reevaluation of the system and goals under which information is classified. DOE has undertaken initiatives to respond to these needs.

DOE Office of Classification. A Classification Policy Study was completed in fiscal year 1992 and is currently in draft form. Its recommendations included the following:

- Redefine restricted data to reflect changes in the security environment, as well as respond to the current state of published information. This recommendation would require amending the Atomic Energy Act.
- Provide authority for the Secretary of Energy to communicate Restricted Data to other countries. Currently, other nations may release information that is, by U.S. standards, Restricted Data and thus may unwittingly aid proliferants.
- Eliminate the Formerly Restricted Data category. Information removed from the Restricted Data category could be protected adequately as National Security Information.
- Allow the authority to reclassify some areas of information. The study found that mandatory declassification of entire areas of information may be too comprehensive. Technological breakthroughs in areas such as special nuclear materials production ought to be classified, for example, but current enrichment techniques should be released.
- Define the scope of Unclassified, Controlled Nuclear Information more precisely.
- Conduct a comprehensive review of all nuclear weapons information, with the objective of removing all information no longer needed to be classified.

DOE Environment, Safety, and Health Initiative. As part of the Secretary of Energy's Safety and Health Initiative (May 1993), DOE was directed to "begin review of Departmental classification procedures and information policies governing public release of documents pertaining to environment, safety and health matters" (33). The review is being directed jointly by the Director of the Office of Intelligence and National Security and the Assistant Secretary for Environment, Safety, and Health. Furthermore, the Secretary directed that "all environment, safety and health documents of the Department of Energy will be prepared with an eye to public release" (33).

SOURCE: U.S. Department of Energy, 1993.

challenge for the Federal Government as a whole, and the specific agencies engaged in these tasks, is to undertake and manage this new post-Cold War mission in a manner that is competent, responsible, and credible. To achieve successful warhead dismantlement and materials management policies and programs in the United States, the Government will need to establish clear policies and well-defined objectives appropriate to present conditions.

The institutions responsible for these tasks must be made equal to the challenge. The Nation will need to establish an institutional structure dedicated to excellence and openness, and to make protection of the environment, safety, and health a working priority in both dismantlement and materials management activities. In addition to a new openness in making and carrying out decisions, effective warhead dismantlement and materials management will require consistent and enduring talent, dedication, and astute management—qualities that government agencies often find difficult to sustain without adequate leadership and vision. Programs and plans will need to be developed through a process that has broad public acceptance, as well as the flexibility to adapt to changing technical and political parameters over the long period during which nuclear materials must be managed.

However, OTA's analyses show that U.S. dismantlement and materials management efforts have lacked focus, direction, and coordination. There has been little informed discussion at high levels of government in terms of planning for the ultimate disposition of special nuclear materials from warheads.

OTA has concluded that the institutions involved in attaining these objectives must meet the following criteria: a management process and culture that give priority to protecting the environment and human health, and promoting safety; internal accountability and external independent regulation or oversight; a mechanism for making information accessible rather than restricted; and a management philosophy of openness, fairness, and public involvement in decisionmaking.

The traditional limitations on access to information about nuclear matters have prevented environmental, health, and safety data from being released and discussed publicly. Lack of relevant information about these factors has heightened public concerns, destroyed public trust and confidence, and increased public opposition to proposed agency actions. If the dismantlement and materials disposition activities are to gain public acceptance, it will be necessary to modify existing limitations on information access and to make data relevant to these aspects of nuclear warhead dismantlement and nuclear materials management easily available to interested citizens.

Further, the ongoing activities and plans with respect to weapons dismantlement and materials management are being conducted without meaningful public involvement. Yet experience from other major nuclear materials management programs—such as the attempt to site repositories for high-level commercial spent fuel, defense transuranic waste, and low-level waste—has shown that when the public, the States, and other affected parties are not effectively included at all stages of relevant deliberations and decisions, proposals by Government agencies are inevitably delayed or derailed. An open, consensus-building process that allows all relevant views to be heard before decisions are made on environmental, health, and safety matters would appear to be essential if key issues (e.g., whether materials from weapons should be used for commercial purposes, where to site nuclear materials storage and processing facilities) are to be resolved with satisfactory and publicly acceptable results.

CONCLUSION

Policies developed entirely behind closed doors are unlikely to achieve public acceptance, particularly decisions that involve significant amounts of Government spending. Public support is necessary for these types of policies to succeed, and public understanding of the issues is a prerequisite for support. Policy development in these areas will depend on the definition of international security in this post-Cold War era—a definition that will inevitably involve not only the role of nuclear weapons but also other concerns relevant to changing conditions. To meet these conditions, a new definition should include broader concepts such as protection of the environment, human health, and safety in a nonmilitary context. For effective policy development, information access will have to be enhanced and participants in the debate will have to come from more sectors of government and society than in the past.

Not only will experts and policymakers at the Departments of Defense, State, and Energy, and the rest of the national security and disarmament communities have to be involved, but the discussion must also bring in the views of those who, in a broader context, have knowledge, authority, and specific interests in protecting human health and safety and preserving the environment. Little will be accomplished unless an informed Nation agrees to pursue common goals regarding nuclear warhead dismantlement, and nuclear materials management and disposition, that preserve the environment, health, and safety.

CHAPTER 5 REFERENCES

1. Adler, Allan R., "Public Access to Nuclear Energy and Weapons Information," *Controlling the Atom in the 21st Century* (Boulder, CO: Westview Press, forthcoming).
2. Advisory Committee on Nuclear Facility Safety to the U.S. Department of Energy, Final Report to the Secretary of Energy, "Final Report on DOE Nuclear Facilities," November 1991.
3. Atomic Energy Act of 1954, 42 U.S.C. Sec. 2011-2296 (1982 and Supp. IV 1986).
4. Broscious, Chuck, Director, Environmental Defense Institute, to Paul Carroll, Office of Technology Assessment, Nov. 13, 1992.
5. Cochran, Thomas, William M. Arkin, Robert S. Norris, and Milton Hoenig, Natural Resources Defense Council, *Nuclear Weapons Databook,* Vol. II (Cambridge, MA: Ballinger Publishing Co., 1987).
6. Costner, Brian, Director, Energy Research Foundation, personal communcation to Paul Carroll, Office of Technology Assessment, Nov. 11, 1992.
7. Costner, Brian, Director, Energy Research Foundation, personal communication, Mar. 30, 1993.
8. Costner, Brian, and Francis Hart, Directors, Energy Research Foundation, Columbia, SC, letter to membership, Jan. 14, 1993.
9. Culp, David, Director, Plutonium Challenge, Washington, DC, personal communication, May 26, 1993.
10. Defense Nuclear Agency, "Description of Nuclear Weapons Dismantlement Process," memorandum prepared for Office of Technology Assessment, November 1992.
11. Department of Energy Organization Act, Public Law 95-91, Title III, Sec. 309.
12. Executive Order 12356 on National Security Information, Apr. 2, 1982, 4 *Federal Register* 14874 (Apr. 6, 1982).
13. Gattis, Beverly, President, Serious Texans Against Nuclear Dumping (STAND), to Emilia Govan, Office of Technology Assessment, Sept. 23, 1992.
14. Gattis, Beverly, Director, Serious Texans Against Nuclear Dumping (STAND), to Jan Linsenmeyer, Office of Technology Assessment, Mar. 6, 1992.
15. Gray, Miles, Chief, Weapons Branch, Office of Classification, U.S. Department of Energy, personal communication, June 2, 1993.
16. Hewlitt, Richard, "The 'Born Classified' Concept in the U.S. Atomic Energy Commission," study prepared for the House Government Information and Individual Rights Subcommittee of the Committee on Government Operations, May 1980.
17. International Atomic Energy Statute, as amended up to June 1, 1973, Article XX, p. 6-5.
18. Martin, Todd, staff researcher, Hanford Education Action League, personal communication to Paul Carroll, Office of Technology Assessment, Dec. 4, 1992.
19. Military Production Network, pamphlet, Washington, DC, March 1992.
20. Napier, John, comments at Office of Technology Assessment, Advisory Panel Meeting, May 5, 1993.
21. National Academy of Sciences, *The Nuclear Weapons Complex--Management for Health, Safety, and the Environment* (Washington, DC: National Academy Press, 1989).
22. Oak Ridge Education Project, *A Citizens Guide to Oak Ridge* (Knoxville, TN: Education Project, May 1992).
23. Olinger, Nancy E., Assistant Attorney General, Environmental Protection Division, State of Texas, personal communication to Kenneth M. Pusateri, Defense Nuclear Facilities Safety Board, Sept. 18, 1992.
24. Presidential Review Directive, National Security Information, Washington, DC, Apr. 26, 1993.
25. Quist, Arvin S., *Security Classification of Information, Vol. 2: Principles for Classification* (Oak Ridge, TN: U.S. Department of Energy, April 1993).
26. Schwartz, Stephen, Washington, DC representative, Military Production Network, personal communication to Paul Carroll, Office of Technology Assessment, June 30, 1992.
27. Schwartz, Stephen, Washington, DC representative, Military Production Network, personal communcation to Regina Hunter, Sandia National Laboratories, June 30, 1992.
28. Secretarial Panel for the Evaluation of Epidemiologic Research Activities for the U.S. Department

of Energy (SPEERA), "Report to the Secretary," March 1990.
29. Serious Texans Against Nuclear Dumping (STAND) of Amarillo, Inc., Information Sheet, 1992.
30. Sinisgalli, A.A., Chief, Programs Operations Branch, Office of Security Affairs, U.S. Department of Energy, personal communication, Apr. 29, 1993.
31. Thomas, Jim, Research Director, Hanford Education Action League, to Paul Carroll, Office of Technology Assessment, Dec. 4, 1992.
32. U.S. Congress, Office of Technology Assessment, *Hazards Ahead: Managing Cleanup Worker Health and Safety at the Nuclear Weapons Complex*, OTA-BP-O-85 (Washington, DC: U.S. Government Printing Office, February 1993).
33. U.S. Department of Energy, "Environment, Safety and Health Initiative," May 1993.
34. U.S. Department of Energy, "Memorandum on Department of Energy Declassification Procedures," May 1, 1990.
35. U.S. Department of Energy, Albuquerque Field Office, "Award Fee Performance Evaluation Report for the period: April 1, 1992-September 30, 1992," Contract DE-AC04-91AL65030, Mason & Hanger-Silas Mason Co., Inc., Pantex Plant.
36. U.S. Department of Energy, Defense Nuclear Facilities Safety Board, *Annual Report 1992* (Washington, DC: U.S. Department of Energy, April 1993).
37. U.S. Department of Energy, Office of Classification, meeting with Office of Technology Assessment, June 23, 1993.
38. U.S. Department of Energy, Office of Environment, Safety, and Health, *Tiger Team Assessments*, various sites and dates.
39. U.S. Department of Energy, Secretary of Energy Advisory Board, "Report of the Task Force on Radioactive Waste Management," 1993.
40. Werner, James, Senior Environmental Engineer, Natural Resources Defense Council, personal communication to Assistant Secretary for Defense Programs, U.S. Department of Energy, Oct. 1, 1990.

Opportunities to Aid Russian Dismantlement 6

The breakup of the Soviet Union and the end of the Cold War present a unique opportunity for the United States and the republics of the former Soviet Union to begin dismantling their nuclear arsenals. Unilateral and bilateral agreements and announcements have formed a basis for both sides to retire weapons systems, destroy delivery vehicles such as missiles, and perhaps make progress in dismantling the nuclear warheads themselves.

The challenge of true mutual and significant reduction of the nuclear stockpile, however, still faces both nations. The United States has begun its own activities to dismantle nuclear warheads, as described in chapters 2 and 3. The extent of corresponding activities in Russia is not clear, but some work is under way.[1] Both nations have made a variety of proposals, and certain agreements are being discussed, but no specific actions have been taken to dispose of nuclear materials from retired warheads. Russia is currently struggling with economic and political problems that may relegate warhead dismantlement and materials disposition to a low priority.

Nonetheless, the United States has expressed its intention to encourage nuclear weapons dismantlement and materials disposal activities in Russia and other former Soviet republics to the maximum extent feasible and has developed its own programs of assistance as a means of helping stockpile reduction become a reality. Congress has provided for a number of recent initiatives aimed at assisting the former Soviet Union to proceed with warhead dismantlement in a safe, secure, and timely manner.

"Among all the huge renewal projects facing Russia today, the main goal is the revival of its industries, including its atomic industry. There are about a million people working for the Ministry of Atomic Energy. We are capable of dismantling up to 5,000 warheads per year. But in order to do this, it is first imperative to undertake and ensure the necessary organization, financing, and provisions for the disposal of nuclear waste."

1992 speech by Viktor Mikhailov, Minister of Atomic Energy of the Russian Federation

[1] Although no U.S. officials have verified warhead dismantlement rates in Russia, some U.S. experts believe that recent statements and evidence indicate a current rate of 1,500 warheads per year (7,11).

This chapter summarizes the information available from unclassified sources about activities, programs, and plans in Russia. It also discusses current U.S. efforts to assist Russian dismantlement and evaluates how well such programs work to reduce future risks from nuclear weapons and materials. Recent progress made in these areas, and prospects for moving forward, are also evaluated.

WARHEAD DISMANTLEMENT IN RUSSIA

The information available on matters related to nuclear weapons activities in the former Soviet Union is very limited, and sometimes conflicting or ambiguous. Publicly available sources indicate that since 1949, the former Soviet Union has produced an estimated 55,000 nuclear warheads (12). Recent statements by the Russian Atomic Energy Minister, Victor Mikhailov, indicate that the Soviet stockpile peaked in 1986 at 45,000 and declined by about 20 percent by mid-1992 (an average of 1,500 per year) (37). Over time, some experts have estimated that Russia has dismantled up to 25,000 warheads, but it is uncertain how much of the material has been recycled into new warheads (12). In various statements over the past year, Russia has indicated that it will retire and "destroy" about 20,000 nuclear weapons,[2] but the exact numbers and types of weapons (and warheads) are subject to speculation.

This chapter focuses on Russia since it is by far the largest republic of the former Soviet Union, and the only one with both the announced intent and the capability to dismantle its own warheads and dispose of the special nuclear materials from those warheads.[3] Russia also has the largest portion of the nuclear arsenal within its borders, compared with weapons now located in other former Soviet republics. Some limited information is available about the situation in these other republics.

The Russian agencies in control of weapons and dismantlement activities are quite similar in function to those in the United States. The Ministry of Atomic Energy (comparable to DOE) has traditionally produced nuclear materials and weapons components, and assembled and tested warheads, while the Ministry of Defense (comparable to DOD) is responsible for weapons staging and stockpile management.

The Russian Ministry of Atomic Energy (MINATOM)—which was created out of the Soviet Ministry of Atomic Power and Industry in January 1992 by decree of President Boris Yeltsin—is responsible for the entire nuclear fuel cycle in Russia, from uranium mining and enrichment to nuclear electricity production and nuclear weapons design, testing, and manufacturing. MINATOM operates nuclear weapons assembly facilities, as well as a number of institutes and laboratories similar in nature to the U.S. system of national laboratories. There are 29 departments within MINATOM, and more than 100 institutes, laboratories, and associations (6,35,40).

The design, testing, and production of fissile materials and nuclear warheads, as well as warhead dismantlement and recycling, have been carried out at about 12 sites. The location of 10 of these sites is so sensitive that they were not marked on any Soviet map and are code named

---

[2] According to certain experts, official documents describing the types of warheads to be eliminated partially or completely include (46): 1) all strategic and tactical warheads withdrawn from Ukraine, Belarus, and Kazakhstan; 2) all nuclear warheads for ground-based tactical missiles, artillery shells, and land mines; 3) half of the tactical bomb inventory of the Air Force, with the remainder removed from frontline units and deployed on bases for centralized stockpiling; 4) one-half of the nuclear warheads for anti-aircraft missiles and one-third of sea-based tactical warheads; and 5) strategic warheads located in Russia, which are to be reduced according to the Strategic Arms Reduction Treaty (START II). The quoted number of warheads to be "destroyed" is 27,000-30,000 (6,54).

[3] The major Soviet Union facilities for assembly and disassembly of nuclear weapons today are in Russia. Three other newly independent states also have nuclear weapons, including Belarus, Kazakhstan, and Ukraine. Officials from Ukraine have stated that they may want to dismantle nuclear weapons located in their country themselves, although it is not clear if they have the resources, knowledge, or facilities to carry this out. There is some question whether Ukraine would violate the Non-Proliferation Treaty if it dismantled nuclear weapons.

after cities 50 to 100 kilometers away. Until recently, the exact locations of these sites were kept secret. Since 1989, most of them have been opened for a limited number of foreign visitors, but others have remained inaccessible. Recently, certain sites have been visited by foreign government and nongovernmental groups, and more information is becoming openly available. Figure 6-1 is a map locating some of the key Russian nuclear weapons sites that have recently been discussed in the open literature.

Information is limited concerning the organization, personnel, and management practices of the Russian Ministry of Defense with regard to nuclear weapons. The Main Administration of Nuclear Weapons (the 12th Main Directorate) of the Ministry is known to be responsible for nuclear weapons staging, storage, and management, once the weapons are obtained from MINATOM. The 12th Main Directorate is apparently responsible for transporting nuclear weapons from MINATOM plants and for staging the weapons at Defense Ministry sites. The Ministry of Defense is also responsible for the transportation of nuclear weapons and warheads from their deployed sites back to MINATOM plants for dismantlement. Apparently, nuclear weapons deployed outside Russia are shipped from their staging sites to central storage facilities in Russia under the control of the Ministry of Defense.[4]

Russian officials have indicated that they are dismantling warheads at the rate of 1,500 to 2,500 per year, but U.S. officials have not verified these dismantlement rates (15). Although definite information about Russian dismantlement progress would be invaluable, it may be difficult to obtain such data without implementing some sort of monitoring to reliably verify the number of warheads going into—and the amount of fissionable materials in storage containers coming out of—a dismantlement site.

ROBERT DEL TREDICI

Partially constructed offices for a breeder reactor complex in the Chelyabinsk region near a Russian plutonium production facility. Construction was stopped in 1991 after a nonbinding public referendum that opposed the building of new reactors.

The lack of knowledge about the former Soviet nuclear arsenal and materials stockpile is recognized as an impediment to international confidence in weapons dismantlement and arms control agreements. During Senate consideration of the first Strategic Arms Reduction Treaty (START) (ratified by the United States on Oct. 1, 1992), a condition to the ratification resolution was added that called for the President to make a good faith effort at negotiating agreements that will allow for the exchange and declaration of information about nuclear weapons and materials stockpiles. Specifically, the condition called for:

> . . . appropriate arrangement(s), including the use of reciprocal inspections, data exchanges, and other cooperative measures, to monitor:
>
> (A) the numbers of nuclear stockpile weapons on the territory of the parties to this Treaty; and
>
> (B) the location an inventory of facilities on the territory of the parties to this treaty capable of producing or processing significant quantities of fissile materials (52).

---

[4] Transportation of most nuclear weapons is by train. At least some nuclear weapons have been transported by train from the staging sites to weapons storage depots in Russia (35).

Figure 6-1—Nuclear Weapons Complex in the Former Soviet Union[a]

[a] Locations are approximate. Facilities are commonly named for the nearest town of significant size.

SOURCE: Defense Intelligence Agency, briefing to the Office of Technology Assessment, Apr. 15, 1993. Graham Allison et al. (eds.), *Cooperative Denuclearization: From Pledges to Deeds* (Cambridge, MA: Harvard University, January 1993). R. Stanley Norris, "The Soviet Nuclear Archipelago," *Arms Control Today*, vol. 21, No. 1, January/February 1992. Thomas Cochran, Senior Scientist, Natural Resources Defense Council, comments, June 1993.

The amendment to the Senate ratification resolution is nonbinding, and does not affect the ratification process for the START II treaty. However, the condition serves to bring warhead dismantlement under the same type of scrutiny that traditionally accompanies arms control verification regimes. Achieving such data exchanges and monitoring arrangements would greatly enhance international confidence in the nuclear nations' dismantlement programs.

Russian warheads most likely are dismantled in the MINATOM plants where they were produced and assembled. Although the Russians clearly possess the facilities for nuclear warhead dismantlement, it is not clear that their economy will be able to maintain the personnel, infrastructure, and financial resources required to operate these facilities.

There are three principal warhead assembly plants in Russia that appear to be similar in function to the U.S. Pantex Plant. They are Sverdlovsk-45, located at Nizhnyaya-Tura (the main facility in the Urals); Penza-19, located at Kuznetsk (115 kilometers east of Penza); and Zlatovst-36, located at Yuryuzan (a smaller facility 85 kilometers southeast of Zlatovst in the Urals). Most sources indicate that these facilities are currently involved in dismantlement activities (6). In addition, the Russian design laboratory known as Arzamas-16 has a small-scale warhead assembly/disassembly capacity and is reported to be involved in warhead dismantlement (see table

Table 6-1—Nuclear Weapons Complex in the Former Soviet Union

Facility name	Activities	Location	U.S. equivalent
Penza-19	Component assembly	Kuznetsk	Pantex/Kansas City
Arzamus-16	Design/assembly	Sarova	LANL/LLNL
Sverdlovsk-45	Assembly	Nizhnyaya Tura	Pantex
Zlatoust-36	Assembly	Yuryuzan	Pantex
Chelyabinsk-70	Design	Kasli	LANL/LLNL
Chelyabinsk-40[a]	Pu/T production	Kyshtym	Hanford
Tomsk-7	Pu/T production/U enrichment	Tomsk	Rocky Flats/Hanford/Oak Ridge
Krasnoyarsk-26	Pu/T production	Dodonova	Hanford
Krasnoyarsk-45	U enrichment	Zernogorsk	Oak Ridge
Sverdlovsk-44[b]	U enrichment	Verkniy Neyvinsk	Oak Ridge
Electrolyzing Chem. Combine	U enrichment	Angarsk	Oak Ridge
Semipalatinsk	Test site[c]	Kazakhstan	Nevada Test Site
Novaya Zemlya	Test site	N. Russia	Nevada Test Site
Ulbinsky Metallurgy	Be/Zr production	Kazakhstan	–

NOTE: Be = beryllium; LANL = Los Alamos National Laboratory; LLNL = Lawrence Livermore National Laboratory; Pu = plutonium; T = tritium; U = uranium; Zr = zirconium.
[a] May also be called Chelyabinsk-65.
[b] Also called Urals Electrochemistry Combine.
[c] Closed 1991.

SOURCE: Office of Technology Assessment, 1993.

6-1 and figure 6-1). In the past, the fissile materials recovered from dismantled warheads were probably recycled into new warheads.

As discussed in chapter 2, the relevant arms reduction treaties mandate only the destruction of the delivery system (e.g., a missile), if that, and say nothing about the fate of the nuclear warhead or the plutonium and other materials contained in the nuclear explosive package. There is speculation that any missiles that have already been removed as part of arms control treaties, along with those missiles returned to Russia from other former Soviet republics, have been stored intact at existing Ministry of Defense storage sites inside Russia, or that they have been only partially disassembled and the warheads are being stored at Ministry of Defense facilities (6,35,40). Up to the present, the United States has not been able to verify the extent of Russian dismantlement of warheads or the subsequent storage of fissile materials.

At the "Third International Workshop on Verified Storage and Destruction of Nuclear Warheads," held in Kiev and Moscow in 1991, a senior arms control adviser to President Yeltsin indicated that nuclear warhead dismantlement is being carried out at two sites at a rate of about 1,500 warheads per year. Although this implies that Russian dismantlement is actually proceeding at this rate, the United States has not confirmed the number of warheads dismantled. Some Russian statements on dismantlement do not make clear whether warheads dismantled in the past are being discussed or whether the materials recovered from these dismantled warheads have been used in new warheads.[5]

According to both the Ministry of Defense and MINATOM, available facilities for the storage of fissile materials recovered from existing warheads are inadequate to store the amounts of plutonium anticipated from current dismantlement plans (8). There are existing facilities for

[5] The available literature does not provide the answer as to what amounts, if any, of fissile materials from warheads have been recycled into new warheads versus the amounts in storage facilities. According to statements by Ministry of Defense and MINATOM officials, however, the most likely scenario is that very little of the fissile material has been stored in the past. Most of it is likely to have been recycled into new warheads.

Box 6-A—Nunn-Lugar Legislation

In 1991 and 1992, Congress authorized $400 million (for a total of $800 million) in the so-called Nunn-Lugar legislation to assist the former Soviet Union in dismantling weapons of mass destruction, including nuclear warheads. Spending required presidential certification that states of the former Soviet Union were adhering to relevant arms control agreements. The actual funds have already been appropriated to the Department of Defense (DOD). Originally, $400 million was set aside in Public Law 102-228, the Conventional Forces in Europe Treaty Implementation Act of 1991, by an amendment introduced by Senators Sam Nunn and Richard Lugar. The specific legislation now in place authorizing this program includes the following:

- The Soviet Nuclear Threat Reduction Act of 1991 (Title II of Public Law 102-228, Dec. 12, 1991) authorized spending $400 million of DOD's FY 1992 budget to "establish a program to assist Soviet weapons destruction." The full name of PL 102-228 is the "Conventional Forces in Europe Treaty Implementation Act of 1991." It is a law dealing with North Atlantic Treaty Organization funding and equipment transfers. Amendment SP 1439, introduced by Senators Nunn and Lugar, became the "Soviet Nuclear Threat Reduction Act of 1991."
- The Soviet Nuclear Threat Reduction Act of 1992 (Title III of Public Law 102-229, Dec. 12, 1991) is a portion of the Dire Emergency Supplemental Appropriations Act of 1992, which makes technical corrections and changes in the budget. It "[a]llows for the transfer of funds to assist the former Soviet Union and/or emerging political structures . . . in dismantling nuclear weapons."
- The Former Soviet Union Demilitarization Act of 1992 (Title XIV of Public Law 102-484, Oct. 23, 1992) sets various conditions for the money authorized under the two bills described above. For example, "defense conversion" money cannot exceed $40 million. It is part of the Defense Authorization Act for FY 1993.
- The Freedom Support Act (Public Law 102-511, Title V, Oct. 24, 1992) provides for economic and nuclear nonproliferation assistance to the states of the former Soviet Union, and authorizes the use of funds made available under specified acts to carry out demilitarization, and economic conversion regarding nuclear weapons. This focuses mostly on repealing Cold War restrictions on trade, etc., with Soviet bloc nations.

SOURCE: Office of Technology Assessment, 1993.

processing recovered plutonium into new warheads. The three plants at which plutonium has been produced—Chelyabinsk-65, Tomsk-7, and Krasnoyarsk-26—are believed to have storage facilities for plutonium, but it is not known whether enough space for additional fissile materials from warheads exists at these sites, or whether there is capacity at any of the sites for the storage of plutonium pits, tritium, or highly enriched uranium.[6]

Views about nuclear warhead dismantlement and plutonium disposition among various Russian organizations are diverse. The positions taken by military, academic, and certain government agencies may be divergent. Dismantlement policy is also a controversial political issue in Russia. MINATOM Minister Victor Mikhailov has become the target of more conservative Russians for his part in promoting Russian warhead dismantlement. Some members of the Supreme Soviet have criticized other activities involving the United States. The possible effect of internal Russian critics on U.S.-Russian cooperation is unknown at this time.

[6] Only the Mayak complex at Chelyabinsk processes spent fuel from power and naval reactors. "Reactor-grade" plutonium is stored there as plutonium dioxide, and between 25 and 30 metric tons may be stored there currently (such material is less than ideal for weapons use but still could be used to make a nuclear bomb). The Tomsk and Krasnoyarsk facilities continue to produce plutonium for weapons. Apparently, those facilities have a related plutonium handling and storage infrastructure (6,12).

U.S. Assistance in Russian Warhead Dismantlement

Since 1991, Congress has enacted several measures providing U.S. assistance for Russian nuclear warhead dismantlement activities (see box 6-A). The Soviet Nuclear Threat Reduction Act of 1991 (Nunn-Lugar; Public Law 102-228, Title II, Dec. 12, 1991) directed the Administration to provide assistance to facilitate nuclear weapons dismantlement and destruction in the former Soviet Union, and authorized $400 million for that purpose. As of March 31, 1993, the Department of Defense had proposed to obligate about $460 million for various purchases and activities related to this initiative that have been agreed to by the United States and Russia or by the United States and other former Soviet republics. Of the total, only $31 million has actually been obligated. The agreements were reached via a series of meetings and exchanges through March 31, 1993, between the relevant agency representatives of these countries (principally Russia) and U.S. agencies.

In the 1992 Dire Emergency Supplemental Appropriations Act (Public Law 102-229), Congress authorized an additional $400 million to be applied for this purpose. The latter provides for the transfer of up to $400 million from DOD operations and maintenance appropriations or working capital account balances to facilitate the transportation, storage, safeguarding, and destruction of nuclear (and other) weapons in the former Soviet Union (13).

DOD is the executive agent for the program and is working closely with the National Security Council, the Department of State, the Arms Control and Disarmament Agency, DOE, and other governmental organizations (13). The Safe and Secure Dismantlement Interagency Steering Group (SSD) coordinates the various activities. The impetus for forming the SSD was concern about the security and control of Soviet nuclear weapons raised after the August 1991 abortive coup d'etat in the Soviet Union (8).

In initial bilateral discussions in Moscow during January 1992, the Russians stated that the greatest impediment to dismantlement was their lack of suitable long-term storage facilities and containers, and inadequate transportation (8). Secretary of State James Baker responded to Russian Foreign Minister Andrei Kozyrev in February 1992 with suggestions covering possible U.S. assistance in transportation and storage, accident response, an accounting and control system, and ultimate disposition of the highly enriched uranium (HEU) and plutonium from warheads (8). In a November 1992 Moscow meeting, the United States reaffirmed an earlier offer to provide aid to expedite the elimination of strategic arms slated for reduction under START II (8). Relevant legislative provisions pertaining to U.S. assistance require the Administration to certify that the former Soviet republics are committed to:

- making substantial investments toward dismantling and destroying weapons; [7]
- forgoing military modernization;
- forgoing the reuse of fissionable materials in new weapons;
- facilitating U.S. verification of weapons destruction;
- complying with all relevant arms control agreements; and
- observing human rights.[8]

The Bush-Yeltsin summit in June 1992 included the signing of four SSD agreements, and discussions in August 1992 led to further U.S.-Russian agreements (8,16). These included:

- an umbrella agreement for providing Nunn-Lugar assistance, naming DOD and MINATOM as executive agents (see appendix C);

---

[7] It is not clear how the Administration is defining "substantial investments" or ascertaining the extent of these investments.

[8] Title II of Public Law 102-228, the Soviet Nuclear Threat Reduction Act of 1991, Dec. 12, 1991.

- an armored blanket agreement;[9]
- an accident response equipment and training agreement;[10]
- a fissile materials container agreement;[11]
- a rail car conversion agreement;
- a storage facilities agreement; and
- an agreement on HEU disposition by conversion to low-enriched uranium (LEU).

In March 1993, Ambassador James Goodby replaced General William Burns as the head of the U.S. Delegation on Safe and Secure Dismantlement of Former Soviet Nuclear Weapons (22). Ambassador Goodby has indicated that the Clinton administration intends to put high priority on agreements under the Nunn-Lugar appropriations to provide incentives for reducing the stockpile and eliminating warheads in the former Soviet Union. The United States is also discussing a possible multilateral approach with other nations such as Japan, the United Kingdom, Germany, Canada, and France. One suggestion is for an international fund with a U.S. contribution through Nunn-Lugar and subsequent appropriations. Each country would lend assistance in its areas of expertise.

Another meeting of the U.S. delegation with its Russian counterparts in Moscow took place at the end of March 1993. During that meeting, the texts of three *new* agreements (22) were agreed upon and await Russian signature under the general umbrella agreement to:

- provide $130 million for equipment to assist with the dismantlement of missile delivery vehicles,[12]
- provide an additional $75 million for special equipment for the planned plutonium storage facility,[13] and
- provide $10 million for improving materials control and accountability systems.

Table 6-2 lists the status of funding for all of these projects as of April 1993.[14]

Different degrees of progress have been made with the other three nuclear states—Ukraine, Belarus, and Kazakhstan—but agreements similar to those with Russia are under discussion for the transportation of nuclear weapons and the dismantlement of delivery systems (8). Efforts to conclude agreements with these states continue—the greatest progress having been made with Belarus, which has already signed agreements (22). Kazakhstan appears to be willing to move forward, but Ukraine has presented some problems (22). The Ukraine Government is now divided over whether it should become, or remain, a nuclear power. The government has stated that it will require $2.8 billion to dismantle its nuclear weapons, whereas the U.S. offer of assistance is in the range of $175 million (22). Little information is available about the breakdown of the $2.8 billion requirement, but U.S. officials generally consider it to be unrealistically high (29).

---

[9] The armored blankets are for the protection of nuclear warheads during transportation. DOD has delivered 250 armored blankets and is sending out for bid procurement of additional blankets (13).

[10] This agreement is to provide emergency response equipment and training to deal with potential nuclear weapons transportation accidents. The United States, through work conducted largely at Sandia National Laboratory, is considered to have the lead in this area. DOD will provide MINATOM with a variety of emergency equipment used for dealing with weapons transportation accidents, as well as initial operator training.

[11] Under this agreement, DOD will provide MINATOM with up to 10,000 containers for exclusive use in transporting fissile materials from dismantled weapons. Design and development were begun by DOE, the containers will be built in the United States, and delivery is anticipated by December 1995 (8).

[12] For the United States, the Defense Nuclear Agency will administer procurement of the equipment, and for Russia, the Committee on Defense Industries will be the executing agency.

[13] MINATOM will be the executing agency for the Russians and the Department of Defense for the United States.

[14] However, in negotiating the sale of Russian HEU, the United States is insisting that Ukraine receive a fair portion of the income from the sale, which may yield $1 billion for Ukraine (22,23).

Table 6-2—Soviet Nuclear Threat Reduction Projects

Recipient nation	Project	Obligations (proposed) as of March 31, 1993 ($ millions)	Obligations (actual) as of March 31, 1993 ($ million)
Russia	Armored blankets[a]	5.0	3.1
	Rail car security[a]	20.0	3.1
	Emergency response[a]	15.0	9.9
	Materials controls	10.0	0
	Storage containers[a]	50.0	2.0
	Facility design[a]	15.0	6.0
	Facility equipment	75.0	0
	Export controls	2.3	0
	Science center[a]	25.0	0.2
	Chemical weapons[a]	25.0	1.9
	SNDV dismantlement[b]	130.0	0
	Military contacts	15.0	0
	Arctic nuclear waste	10.0	0
Subtotal		**397.3**	**26.4**
Ukraine	Emergency response	5.0	0
	Communications	2.4	0
	Export controls	2.3	0
	Materials controls	7.5	0
	Science center	10.0	0
Subtotal		**27.2**	**0**
Belarus	Emergency response[a]	5.0	3.3
	Communications[a]	2.3	0
	Export controls[a]	2.3	0
Subtotal		**9.6**	**3.3**
Kazahkstan	Emergency response	5.0	0
	Communications	2.3	0
	Export controls	2.3	0
	Materials controls	5.0	0
Subtotal		**14.4**	**0**
General	Support/assessment	10.0	1.6
TOTAL		**458.7**	**31.3**

[a] Denotes signed agreements.
[b] SNDV = Strategic nuclear delivery vehicles.

SOURCE: U.S. Department of State, 1993.

The people of Muslyumovo from the Chelyabinsk region. Muslyumovo is less than 50 miles downriver from a plutonium processing plant that dumped high-level radioactive waste into the Techa river from 1948 to 1951.

Almost all of the assistance to Russia for weapons dismantlement is in either of two forms: 1) supplies or equipment purchased in the United States, or from U.S. stockpiles, to be shipped to appropriate Russian agencies; or 2) U.S. technical or advisory teams to supply technical services or data to appropriate agencies. Russia has not yet sought any assistance for actual weapons dismantlement, and it opposes U.S. involvement in such activities.[15] It appears that Russia is concerned the United States would gain too much access to secret facilities or information, and Russia is confident of its own dismantlement capabilities. To date, all U.S. assistance has been based on funds spent in the United States for goods or services to be delivered later to Russian agencies.

In each case, U.S. negotiators have tried to determine the need for assistance and assess its importance before reaching an agreement with the Russians. Russia has expressed a number of needs and pressed for the direct commitment of funds for building facilities. However, the U.S. policy is to retain essentially complete control over spending of funds under Nunn-Lugar and to purchase services or materials in the United States, consistent with the intent of the Nunn-Lugar legislation to maximize the use of U.S. technologies and technicians (14). The program is administered by DOD, and most of the purchases of services or equipment are managed by the Defense Nuclear Agency. The Army Corps of Engineers is, at present, executing only the

---

[15] According to the *Weapons Complex Monitor* (55): ''General Sergei Zelenstov, chief nuclear engineer of the Russian Ministry of Defense, said all tactical nuclear weapons in former Soviet republics except the Ukraine and Byelorussia have been removed. A top Russian official said 'We don't need any technology or technical aid for dismantling warheads. We don't need a United States contractor. We can do it ourselves.' Russia has however asked for aid to develop storage facilities.''

storage facility design (14). This policy could be reexamined if U.S. goals for Russian dismantlement are not achieved.

Although Russian President Yeltsin supports the SSD program and work now under way, the Supreme Soviet has criticized it on the grounds that it diminishes Russia's status as a nuclear superpower to have the United States dictate specific aspects of its nuclear enterprise. This, in turn, has led President Yeltsin to keep a low profile on the SSD program for the time being and also to postpone submitting the START II treaty for ratification (22). In addition to some dissatisfaction with what are viewed as U.S. efforts to unilaterally dictate the nuclear policy of former Soviet Union nations, officials from both Russia and Ukraine have expressed frustration with the slow pace at which U.S. dismantlement aid is being made available (42).

Even in Russia, the political climate for acceptance of U.S. assistance is problematic. The current program is a compromise between the two countries. Agreements have been reached slowly and in small steps. Some critics believe that expenditures have not been well targeted and have not led to significantly safer or quicker dismantlement (11). It does not appear that this program is being used to address the broader issues of mutual goals and interests in the overall weapons dismantlement programs of both countries. The timing of U.S. efforts to address these issues may become more critical if political and economic instability in Russia continue (9).

MANAGEMENT OF RUSSIAN PLUTONIUM

Little information is available on how the former Soviet Union managed plutonium from dismantled nuclear warheads in the past, including how long and where it may have been stored as intact pits, and where and how it was reprocessed or recycled into new warheads. Some experts believe that plutonium was fabricated rapidly into new warheads so that large storage facilities were not required.[16]

There are conflicting reports on whether Russia currently has sufficient storage capacity to carry out dismantlement activities.[17] Reports from some meetings in Russia indicate that lack of a dedicated storage facility will not delay dismantlement because temporary storage facilities are available. The name or location of these facilities is unclear (12). On the other hand, Russian officials assert that lack of containers and storage space for highly enriched uranium and plutonium is the limiting factor in Russia's dismantlement effort[18] (8). This contradiction indicates that more specific information about Russian capabilities and facilities may be needed if the assistance program is to be effective.

U.S. Assistance to Russia for a Plutonium Storage Facility

As is the case with U.S. weapons, the nuclear materials from dismantled warheads of the former Soviet Union will have to be stored until a decision is reached on final disposition. The United States is providing assistance in the design of a storage facility for nuclear materials recovered from dismantled Russian warheads (via the origi-

[16] In the United States, plutonium pits are commonly reprocessed to remove impurities (americium-241 in particular) prior to being remade into new pits. According to one source, both the Soviet Union and the United States recycled plutonium recovered from dismantled warheads into new warheads, and therefore did not generate large plutonium stocks until the late 1980s (53).

[17] The three plants, described above, in the former Soviet Union at which plutonium was produced have stored at least small amounts. At the end of 1991, 25 metric tons of separated plutonium in oxide form from civilian reactors was stored at the Chelyabinsk-65 plutonium production facility (12).

[18] A Russian news report dated August 6, 1992, stated that "the nuclear workers themselves attest that Krasnoyarsk-26 [another plutonium production site] is joining the process of dismantling nuclear warheads. The tons of plutonium produced here will most likely also return here—for storage" (25). Another former plutonium production facility—Tomsk-7—is the site currently being considered for construction of the large storage facility for fissile materials from retired warheads.

nal Nunn-Lugar Soviet Nuclear Threat Reduction Act, Public Law 102-228, and subsequent laws).[19]

An agreement between the U.S. DOD and Russian MINATOM to assist Russia in the safe, secure storage of fissile materials from dismantled weapons was announced in August 1992. It committed DOD to assist a Russian-led effort in designing a storage facility. Initial Russian design requirements were received in August, and several joint technical meetings have been held. The United States has had ongoing discussions with Russia about construction of a storage facility there for nuclear materials from dismantled warheads.

DOD is committed to provide, at no cost to MINATOM, technical assistance in this design effort, including the development of design requirements and criteria for the storage facility. Under this agreement, MINATOM retains full responsibility for facility design, and the total DOD cost may not exceed $15 million (1). This amount is considered to be only for design assistance and is not expected to cover any fraction of actual construction. The total construction cost and the source of funding have not been established. Although the United States would clearly like Russia to cover the major portion of these expenses, Russia may be unable to provide the amounts needed.

The Army Corps of Engineers is responsible for storage facility design assistance under the agreement.[20] DOE is providing design expertise from the Albuquerque Operations Office for project oversight; the Los Alamos National Laboratory (LANL) for control, accounting, and safeguards; and the Sandia National Laboratory for physical security and materials handling (48,49). The design is intended as a joint U.S.-Russian project, and by March 1993, several U.S.-Russian technical workshops had been held. For example, in December 1992, the Russian delegation, the Corps, and LANL staff drafted general safety criteria (20). The U.S. team is also assisting the Russians in a preliminary safety analysis, although the United States ultimately considers any analysis of environmental impact to be Russia's responsibility (48,49). The scheduled date for completion of the design remains December 1993, although there is uncertainty whether the Russians will be able to complete all their design tasks by that date (14).

Although the Russian fissile materials storage facility design is proceeding with U.S. assistance, the completion date for construction has been delayed 1 year, to 1997 (22). The Russians plan to locate the storage facility at Tomsk-7, near the plutonium production facility. When the new facility was announced, however, the population in the adjacent community registered some opposition (8). In light of the accident at Tomsk-7 in early 1993, when one tank at a weapons material reprocessing plant exploded and released radioactive material, such opposition may reappear (22). The city council of Tomsk voted against locating the plutonium storage facility in that region, and President Yeltsin has stated that these desires would be respected (53). Like the United States, Russia plans to store the plutonium initially as weapons pits, but these may be removed from storage later for further use or converted to a more stable form for longer-term storage. Many Russian officials consider storage to be mainly an interim measure that is required while final disposition plans and technologies are being developed. There is general agreement among government officials that excess plutonium should be used eventually in power-producing reactors, but that these technologies will require considerable investment to be operational.

[19] The Department of Defense plans to obligate up to $90 million to support the storage facility's design, construction, and outfitting (27); $15 million of the $90 million can be used for facility design (8). Up to $75 million is proposed to provide the specialized equipment necessary to make the storage facility operational (13).

[20] The Transatlantic Division of the Corps in Winchester, Virginia manages the program.

Russian officials have agreed in principle that the United States will be able to monitor the fissile materials storage facility. One condition of U.S. assistance under the Nunn-Lugar legislation is that fissionable materials from dismantled Russian warheads will not be reused in weapons. The details of certification are not clear. One of the U.S.-Russian agreements covers the development of a materials and accounting security system (probably with assistance from LANL) for use throughout Russian nuclear operations. By helping to design these systems, the United States hopes to minimize proliferation concerns (8).

U.S. AGREEMENT FOR PURCHASE OF RUSSIAN WEAPONS HEU

The United States and Russia signed an agreement on February 18, 1993, to convert 500 tons of HEU from Russian warheads to LEU and sell it to the United States (see appendix C). This agreement calls for the safe and prompt disposition of HEU from dismantled Russian weapons by conversion "as soon as practical" to LEU [21] for use in commercial nuclear power reactors. Executive agents for the agreement are DOE (or the successor U.S. Enrichment Corporation) and the Ministry of the Russian Federation of Atomic Energy. HEU is a comparatively greater security concern than plutonium because it is technically much easier to construct a bomb from HEU. On the other hand, HEU is easier to "de-weaponize" by dilution with unenriched uranium to 2 to 5 percent uranium-235 used in civilian power reactors (34). Uranium with less than 20 percent enrichment is not considered to be weapons grade.

The HEU agreement specifies that it will take place "in accordance with existing agreements in arms control and disarmament" and to further the objectives of the Non-Proliferation Treaty (NPT), and will "comply with all applicable non-proliferation, physical protection, nuclear material accounting and control, and environmental requirements." The agreement provides that both Russia and the United States are to maintain physical protection of the HEU and LEU, and to implement the relevant International Atomic Energy Agency (IAEA) recommendations and NPT safeguards. Each country would also establish transparency measures, including provisions for nuclear materials accounting, control, and access, from the time HEU is made available until it has been converted to LEU (2).

To carry out the agreement, the United States has established an interagency task force that includes the National Security Council (which has the lead), the State Department, the Arms Control and Disarmament Agency, the Office of Management and Budget, DOE (Nuclear Energy Office), DOD, and the Nuclear Regulatory Commission (18,43). The agreement covers the amounts of LEU to be purchased, the security necessary, and the location at which conversion will occur. It also provides for participation of the private sector in the U.S.-Russian enterprises.

The State Department is continuing to negotiate the U.S.-Russian HEU purchase agreement. It considers the basic terms to have been established, but two issues remain to be negotiated: transparency and how profits will be divided between the four nuclear states of the former Soviet Union (47). The allocation of profits from the sale is a major difficulty that is yet to be resolved.

Quantities, Location, and Transportation of Russian HEU

The HEU agreement specifies that approximately 500 tons of Russian HEU from dismantled warheads is to be converted to LEU in Russia.[22] Most experts believe that this represents a substantial proportion of the HEU to be recovered from dismantled Russian warheads, although the

21 Defined as uranium enriched to less than 20 percent concentration of the isotope uranium-235.

22 In all cases, tons as used in this chapter refers to metric tons.

Box 6-B—U.S. Facilities for Handling Russian Weapons Uranium

If and when LEU is shipped to the United States from Russia, certain private fuel fabricators such as General Electric, Westinghouse, Siemens, or ABB (see figure 6-2), could be engaged to convert it to a usable product for commercial reactors. They do not have licenses to store or handle uranium above 5 percent enrichment but could process LEU to exact customer specifications if Russian HEU were blended down to less than 5 percent uranium-235 before shipment.

Another option would be for Russia to blend HEU to less than 20 percent enrichment. At this level, two private companies, Nuclear Fuel Services (NFS) in Erwin, Tennessee and B&W Corporation in Lynchburg, Virginia, as well as Department of Energy facilities at Y-12 and Portsmouth, could handle the more highly enriched material.

If HEU were shipped directly from Russia (as provided for, in principle, in the agreement), the basic technology for blending HEU into LEU is available here (32). Examples of small-scale U.S. blending facilities include the NFS facility that has processed and delivered HEU naval reactor fuel under a Nuclear Regulatory Commission license (4,43). B&W Corp. also has a naval fuel fabrication facility in Lynchburg that is similar, but smaller (18). Allied Signal has processed uranium concentrates into uranium hexafluoride at its Illinois plant (4). At NFS and B&W, however, the capacity to convert uranium to uranium hexafluoride would have to be added. Although these technologies have been used on a small scale, they have not yet been employed on the scale required if the United States were to import Russian HEU.

SOURCE: Office of Technology Assessment, 1993.

precise proportion is subject to speculation. By August 1993, an implementing contract is to provide for the conversion of no less than 10 tons of HEU per year for five years—followed by 30 tons per year—requiring a total of 20 years for the conversion of 500 tons. When converted to LEU, 500 hundred tons of HEU would provide enough fuel to supply all U.S. requirements for about 10 years (19).

Russian officials have stated that they have the necessary capacity to blend the Russian HEU to LEU (U.S. facilities are described in box 6-B and figure 6-2). A technical team from the U.S. uranium companies Allied Signal and Nuclear Fuel Services (NFS) toured a Russian conversion facility, Sverdlovsk-44, at Verkniy Neyuinsk, near Yekatrinburg. They reported that new facilities are being installed at which HEU will be converted and blended with uranium hexafluoride (UF_6) to produce LEU in the form of UF_6. In addition, the Russians are installing loading facilities for UF_6 shipping containers (44). The UF_6 fluorination facility is expected to have a test run in October 1993. DOE's Office of Nuclear Energy also believes that the Russian facilities at Sverdlovsk-44 will be adequate for converting and blending Russian HEU to LEU hexafluoride (45). Although the Office of Nuclear Energy thinks that the results of the Allied Signal-NFS review support this viewpoint, it is conducting its own investigation and tour. Some modifications and additions to this facility may be required to achieve the maximum capacity of 30 tons per year stated in the agreement. There is also some question about the quality of Russian HEU relative to U.S. standards for fuel fabrication (see box 6-C), and the blending operations may have to accommodate processes to ensure quality.

Under the purchase agreement, LEU, after conversion, would be shipped in the form of UF_6 in commercial shipping cylinders. DOE expects to receive these shipments at its existing enrichment facilities (Portsmouth and Paducah) for the purpose of final blending to meet private customer specifications. Alternatively, blending could be done by private firms under arrangements with both DOE and Russian parties. After blending, the LEU would be shipped to a fuel fabricator for conversion to an oxide form and the manufacture of fuel rods (see figure 6-2).

Figure 6-2—Proposed Facilities for Handling and Processing Uranium from Russian Warheads

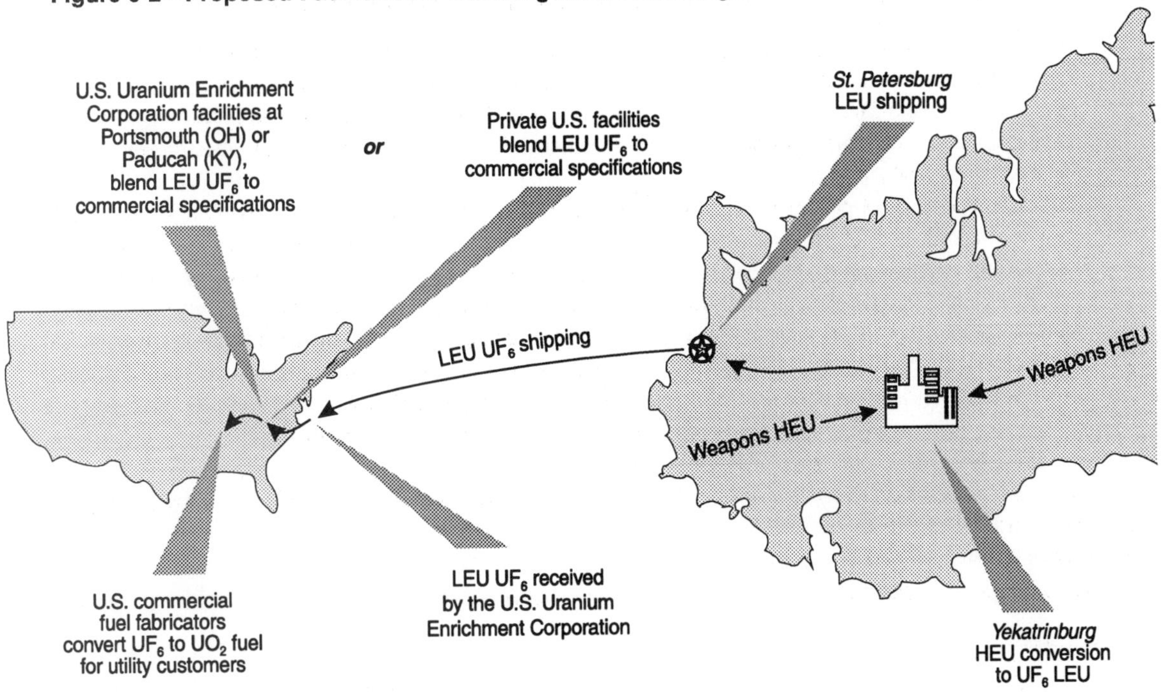

SOURCE: Office of Technology Assessment, 1993.

The agreement of February 18, 1993, states that "... an equivalent amount of HEU can substitute for the corresponding amount of LEU planned for purchase by the U.S." If HEU were transported from Russia, the cost and difficulty would escalate because of required security measures. Regulatory issues would also have to be considered if HEU blending occurred in the United States. The Nuclear Regulatory Commission oversees HEU and LEU processing in the private sector (although not at DOE Weapons Complex facilities). It issues licenses to store, transport, and process nuclear materials. Regulations governing private U.S. processing facilities are being revised in response to health and safety problems (including near-criticality conditions) (38,39). Commercial nuclear facilities are also subject to regulation under the Environmental Protection Agency (EPA) and the Occupational Safety and Health Administration (OSHA). Portions of DOE and private-sector facilities that process HEU are subject to security measures that could complicate IAEA inspection procedures. Currently DOE facilities are not subject to the same EPA and OSHA requirements that apply to commercial facilities (see chapter 2).

The Russian-U.S. HEU agreement, if implemented as planned, could have consequences for world security, economic matters, and world peace, including the following:

- Dilution to LEU might reduce the threat of nuclear proliferation since LEU can neither be used directly to make nuclear weapons nor be turned back easily into HEU.
- Revenues from the sale of HEU might bolster the economies and social stability of former Soviet Union states; fund dismantlement activities; and support urgently needed environmental restoration and health and safety measures at their weapons facilities and nuclear reactors.

Box 6-C—Problems with Russian HEU Quality

The quality of Russian HEU available under the agreement may be a problem because of its contamination with undesirable uranium isotopes (18). The uranium in LEU fuel is predominantly uranium-238, with lesser amounts (e.g., 3 to 5 percent) of uranium-235. Other uranium isotopes such as uranium-232, uranium-234, and uranium-236 are considered undesirable: uranium-234, an alpha emitter, is regulated because of the occupational risk it poses (18, 36). An American Society for Testing and Materials specification limits the maximum content of uranium-234 in fuel to 1 percent (18).

Concentrations of these undesirable uranium isotopes relative to uranium-235 can be increased either during normal uranium enrichment processes or during the reprocessing and recovery of uranium in spent nuclear fuel (18). Uranium-234 is a naturally occurring isotope that becomes enriched along with uranium-235 during the normal enrichment process. In addition, uranium-234 fissions more slowly than uranium-235 during irradiation in a nuclear reactor (i.e., uranium-235 is used up more quickly). Thus, reprocessed uranium recovered from irradiated spent fuel is even more enriched in uranium-234. The Russians have mixed and recycled their civilian power reactor and military uranium, and their HEU may now be contaminated with these isotopes at levels that are unacceptable by U.S. standards (18).

Minimally contaminated uranium might be cut with depleted or enriched uranium to produce an acceptable product. However, if Russian HEU cannot be blended to below this level of undesirable isotopes, U.S. fuel fabricators may be reluctant to accept it (18).

SOURCES: U.S. Department of Energy and Office of Technology Assessment, 1993.

Security Enhancement from the U.S.-Russian HEU Agreement

A major incentive for U.S. purchase of Russian weapons HEU is to limit the security and proliferation threat represented by this material as long as it remains in Russia. Some consider that the most effective method for preventing proliferation is to limit, as much as possible, access to special nuclear materials (26). Although the relative value of HEU and plutonium (to a reasonably technically advanced nation or group) probably depends on how readily obtainable each is, rather than how it would be used in a weapon, HEU may be more attractive in some ways. HEU, but not plutonium, can be used in a "gun-type" weapon, which would be easier to design and would have a greater chance of working than a bomb based on "implosion" (33). This advantage would be especially appealing to a state with limited technical capability or to a subnational terrorist group. On the other hand, the gun design requires substantially more nuclear material. Therefore, HEU may be more attractive than plutonium to those interested in certain potential weapons if enough material could be obtained.

The stated rationale for the HEU agreement is that it will enhance security and reduce proliferation potential. President Bush announced that the agreement was intended to ensure that Russian HEU from dismantled nuclear weapons would be used exclusively for nonmilitary purposes via conversion to civilian reactor fuel, and that it established nonproliferation, physical security, materials accounting and control, and environmental requirements (56). Intuitively, the HEU agreement appears beneficial for both U.S. and world security, but little analysis is available because of a lack of relevant information.

For example, without knowing the current and potential Russian HEU inventory, it is difficult to fully assess the security value of the agreement. Very little information is available about Russian uranium inventories, production capabilities, and practices (51). MINATOM supervises the entire chain of production and use of nuclear materials in Russia, including mining of uranium ore, enrichment, construction of reactors, and waste management (4). In the late 1940s the Soviet Union began enriching uranium for weapons, and by the 1980s it had gained a reputation as a dependable supplier of enrichment services in Europe (51). DOE projects that Russia will have a significantly larger share of the U.S. market for enriched uranium, the largest uranium market in the world, by the mid- to late 1990s and that, by the end of 1992, Russia will obtain 65 percent of all new contracts (51). In 1991, the Soviet Union operated four gas centrifuge enrichment plants (the U.S. uranium enrichment industry is based on gaseous diffusion enrichment), which according to a Ministry of Atomic Power and Industry official in 1990 were all located in Russia (51). An industry report estimated that Russia's total capacity is at least 10 million SWUs (separative work units) per year, which is more than the total U.S. annual consumption of uranium fuel for commercial reactors (51). Russian uranium enrichment capacity is presumably the same today as it was before the breakup of the Soviet Union. With Russian interest in exporting its enrichment services to earn needed hard currency, it may be motivated to continue enrichment operations.

The timetable of the HEU agreement requires 20 years for 500 tons of Russian HEU to become de-weaponized by conversion to LEU, but it is not clear what proportion of the total Russian HEU stockpile this represents. The 500 metric tons was a figure volunteered by Russia about how much material would be made available after dismantlement (45). Although the U.S. State Department estimated that scheduled nuclear warhead dismantlement in Russia would generate about 500 tons of HEU, it recognized the possibility of other HEU stockpiles not contained in warheads (34). Furthermore, HEU may remain in warheads not scheduled to be dismantled, so it is not known whether all Russian HEU would be converted to LEU.

However, DOE considers that converting any amount of HEU from Russian stockpiles is a positive step for U.S. security and world peace (45). Yet, without knowing the Russian plans for conversion to LEU, it is difficult to assess the impact of the agreement on the amount of HEU transported in Russia and the corresponding risk of diversion or theft. Russian HEU may now be located at one or more storage facilities. During or after the 20-year period when, under the agreement, Russian HEU stocks are to be drawn down, it is not clear to what extent the risk of theft will be reduced.

Russia retains the capability to produce more HEU. Although Russia, like the United States, has not produced any HEU for some years (8), the agreement does not address the issue of possible future HEU production in Russia. Instead, the agreement is based on the premise that the uranium purchased by the United States would come from dismantled weapons.

Critics have suggested that the agreement to purchase 500 tons of HEU may be more of a symbolic than a practical measure, which will have little impact on reducing the threat of nuclear weapons at least in the near term. According to this perspective, the agreement may be primarily a way to aid the Russian economy, with only a marginal contribution toward reducing the threat of nuclear weapons (21). On the other hand, bolstering the economy and social structure of the republics of the former Soviet Union through this means may have its own security rewards.

Also to be considered are the practical realities of implementing the provisions regarding compliance with IAEA safeguards and materials protection standards, although actual IAEA enforcement of these standards is *not* contemplated in the agreement. Even if the parties agree in the future to an IAEA role, it may be difficult to implement. Although the IAEA has the legal authority to take and store excess fissionable materials (relative to the amount required for civilian use) including uranium, it may not have the necessary resources (34). The IAEA may continue to have funding problems that will preclude any increased role in security and verification (10), unless funded by the United States and other nations.

Economic Benefits from the U.S.-Russian HEU Agreement

By most criteria, the incentive to achieve enhanced security outweighs most considerations of economics or profit. Yet, many within DOE and U.S. private industry have looked at the U.S.-Russian HEU agreement primarily as a potentially profitable business deal, although it is to be budget neutral for the U.S. Government on a year-to-year basis.[23] Russia also has positive financial expectations since the agreement specifies that it is to use some proceeds for the "conversion of defense enterprises, enhancing the safety of nuclear power plants, environmental clean up of polluted areas and the construction and operation of facilities in Russia" (2).

Processing and diluting HEU to LEU in Russia could maintain Russian jobs at defense-related facilities that might otherwise be downsized or closed. Stabilizing the Russian economy may be crucial in maintaining sufficient political stability for the Russian nuclear warhead dismantlement program to proceed as hoped. The Russian Ministry of Atomic Energy has more than a million employees. However, the United States will have to be convinced that maintaining employment at Russian defense facilities would not in effect invigorate the Russian nuclear weapons complex. The Nunn-Lugar provisions partially address this point by requiring that nuclear materials recovered from dismantled Russian warheads and stored in facilities built with U.S. assistance must be certified not to be reused in new weapons.

Victor Mikhailov, the Russian Minister of Atomic Energy, has stated that Russia would invest "hundreds of millions of dollars" in profits from the sale of HEU into cleaning up its environment and building safer nuclear power plants and facilities (31). DOE and the U.S. State Department consider that the Russians are motivated primarily by the opportunity to earn hard, Western currency from their sales of LEU to DOE (18,34).

DOE perceives the agreement as an opportunity for Russian uranium enrichment operations to prosper in the competitive worldwide nuclear fuel business. Although in 1969 the United States had a monopoly over the Western World's uranium enrichment market, foreign investors had taken over most of the struggling U.S. uranium mining industry by 1988, and imports supplied 51 percent of U.S. power utility requirements (50). In 1992, DOE's share of the world enrichment services market was reduced to 40 percent (24,28). Sales of the 500 tons of Russian HEU after conversion to LEU would be equivalent to about 7 years of DOE's enrichment services (43) or, spread over 20 years, to an average of 35-40 percent of DOE's enrichment sales annually (32). The key economic benefit for DOE may be to reduce operating costs and thus remain a competitive source of uranium fuel in the next century (32).

No price was indicated in the February agreement. Current price negotiations will be announced when all terms and conditions have been agreed to. A key means of valuing Russian HEU is the health of the uranium market for the remainder of this decade. Some economic forecasts for the nuclear power industry are not optimistic. In an analysis of the installed nuclear-generating capacity worldwide, the resulting uranium demand, and the interaction between demand and supply, the conclusion was that installed capacity is likely to increase very slowly to the year 2000 and that there is little likelihood of substantial real increases in uranium prices (41). Past government and industry forecasts of installed nuclear-generating capacity have been consistently overoptimistic. Most forecasts were compiled by agencies that are strongly committed to nuclear development and did not adequately

[23] In addition to the goal that the agreement be budget neutral, it will not be funded under the Soviet Nuclear Threat Reduction (Nunn-Lugar) Acts of 1991 and 1992.

recognize the potential for delays or cancellations arising from engineering, bureaucratic, and political obstacles.

Recent legislation that will convert DOE's Uranium Enrichment Corporation into a semiautonomous government corporation (the Energy Act of 1992, Public Law 102-486) may have some impact on the Russian HEU agreement. Title IX of the Energy Act, which established the U.S. Enrichment Corporation (USEC), also deals explicitly with the purchase of HEU from the former Soviet Union (section 1408). The law authorizes USEC to negotiate the purchase of all HEU made available by any state of the former Soviet Union under a government-to-government agreement, provided that the "quality of material can be made suitable for use in commercial reactors." Whatever plans are adopted, in the early years of the agreement the USEC is expected to use HEU-derived LEU to satisfy contracts with utilities. The question of budget neutrality over the long run is effectively moot once USEC takes over as executive agent of the agreement since it will no longer be an agency of the U.S. Government and will have to pay its own way. Nevertheless, the long-term financial risk is that the U.S. Treasury will eventually have to cover any losses suffered by USEC or that it will have to guarantee a loan to pay for the uranium on which USEC could default.

Potential Obstacles to the U.S.-Russian HEU Agreement

The Clinton administration has continued negotiating the implementation contracts (originally to be signed within a few months) specified in the U.S.-Russian HEU umbrella agreement (23). Several issues remain to be resolved.

For example, the United States required that the basis for dividing profits from HEU sales among the former Soviet republics involved (Russia, Ukraine, Belarus, and Kazakhstan) be established at the time of the agreement. Russia and Ukraine have not been able to agree on this.

Also to be negotiated are specific provisions for U.S. verification of the weapons source of HEU, as well as other materials control and accountability procedures. In addition, the availability of HEU from dismantled Russian weapons depends on Russia's capabilities to maintain a dismantlement schedule that it states is subject to the availability of a new storage facility.

It is unclear when these difficulties will be resolved, but both parties are working on them. Another obstacle may be the concern about Russian uranium replacing U.S. supplies. Uranium mining and enrichment interests in the United States have objected to the sale of HEU and linked this to complaints about Russia's alleged uranium dumping (see box 6-D). Unions representing workers at uranium enrichment facilities are concerned that importing this Russian uranium will displace U.S. jobs at these plants. An October 1992 uranium antidumping accord exempts Russian HEU from quotas (1), but the antidumping agreement remains contentious.

CONCLUSION

Although the United States has initiated a number of efforts to aid Russian dismantlement that are important first steps, concrete benefits from these efforts may be limited. This limitation may result in part because no adequate strategic analysis has yet been carried out of the most important immediate and long-term objectives of the United States and Russia with respect to warhead dismantlement and materials management, nor is there a plan for attempting to reconcile differing national objectives and requirements.

One issue for consideration is whether the weapons dismantlement assistance that the United States is willing to provide coincides with the priority needs of Russia. Although dismantlement of weapons is a matter that the United States views as important for international security purposes, the economic and political situation in Russia may make it difficult for resources to be

Box 6-D—Uranium Antidumping Suit

How can HEU from nuclear weapons dismantlement be converted and sold on the commercial uranium fuel market without further upsetting the already depressed U.S. uranium mining and enrichment industry? When the previous Administration announced the HEU agreement, it claimed that it would have no adverse impact on U.S. consumers or jobs in the uranium mining, or processing industries (56). If Russian weapons uranium were blended with newly mined U.S. uranium, the impact on mining jobs might be minimal. However, the current approach to let most blending be done in Russia would, undoubtedly, have the effect of replacing U.S. uranium supplies, unless DOE continues to purchase and stockpile natural uranium.

The uranium miner's union has objected in the past to the likely impact of the importation of Russian uranium (not from weapons) to the United States, and it brought suit to block Russian commercial sales. In November 1991 the Ad Hoc Committee of Domestic Uranium Producers, a coalition of 13 U.S. mining and milling companies and the Oil, Chemical and Atomic Workers International Union, whose members are uranium conversion and enrichment plant workers, filed an antidumping petition with the Commerce Department's International Trade Commission and International Trade Administration. The petition requested relief under the Tariff Act of 1930 (19 U.S.C. 1673a) (50). It claimed that members had been injured by the sale of Soviet uranium at less than fair market value (50). In 1992, as a result of this suit, the U.S. Commerce Department found that Russia had been selling commercial-grade (not from weapons) LEU at unfairly low prices (31).

This case has so far addressed only the dumping of commercial-grade Russian LEU and has excluded Russian weapons uranium from consideration (17). Although the Department of Commerce determined that Russia was indeed dumping uranium, it specifically excluded HEU from its preliminary ruling, which implies that Russian weapons uranium may be freely imported into the United States (30, 32). The Uranium Miner's Union may object to this ruling (32).

Agreements ending the U.S. investigation of alleged dumping of uranium by the states of the former Soviet Union were signed on October 16, 1992 (3). The agreement with the Russian Federation states that the Russian Ministry of Atomic Energy will restrict the volume of direct and indirect exports of uranium products from all sources in the Federation. HEU in existence at the signing of the agreement, or any LEU derived from it, is exempted from the quotas imposed on uranium shipments from Russia, provided it is purchased by the Department of Energy or U.S. Enrichment Corporation. Also, the Russian Federation is granted a one-time only opportunity to sell a fixed quantity of uranium through the end of 1994. The exemption of HEU in the antidumping agreement is explicitly linked to the agreement on the purchase of HEU.

It is unclear whether any future challenges to this agreement may be forthcoming from the U.S. uranium production industry or the unions. The Nation must weigh the security benefits of purchasing Russian HEU against any job impacts that may occur.

SOURCE: Office of Technology Assessment, 1993.

devoted to that purpose, and current U.S. assistance efforts alone may not ensure that Russia's dismantlement process will move forward. If U.S. efforts do not address some critical needs and interests of the former Soviet Union along (and perhaps connected) with dismantlement matters, little may be accomplished in terms of the latter. By continuing to address these issues separately in dealings with Russia, the United States may not be able to achieve optimal results.

Another important issue is whether any storage or processing facilities used in connection with warhead dismantlement and materials management should be subject to international monitoring, inspection, or even control. In that regard, it remains to be seen whether the United States can realistically expect to verify, either directly or through international agencies, that Russia is proceeding with safe storage of special nuclear materials, without some reciprocal interest by

Russia in verifying U.S. progress along the same lines. At present, there does not appear to be a high-level governmental process to consider and address such issues.

The ongoing initiatives and activities to assist Russia are more likely to obtain good results if the United States develops a more focused and integrated program within the context of broad policy objectives. Whether or not Russia calls for reciprocal actions by the United States, many believe that global nuclear arms reduction will come about only if the United States sets an example through its own warhead dismantlement and materials management policies and programs (5). Thus, the efficacy of any U.S. attempts to influence Russia could be enhanced to the extent that what is expected of Russia has some relationship to the actions of the United States itself. In formulating this overall program, leaders will have to understand what changes in U.S. programs could result in corresponding changes in Russia's program.

The current governmental process to consider and address such issues has not yet resolved them. In addition, U.S. Government efforts with respect to Russian weapons dismantlement and materials disposition have not always been well coordinated. Since various offices and agencies are dealing with different portions of the initiatives, the essential linkages and connections among the initiatives are not always analyzed or considered. This could eventually lead to problems in effectively implementing existing programs or developing additional ones.

CHAPTER 6 REFERENCES

1. Agreement between the Department of Defense and MINATOM Concerning Technical Assistance for Design of a Safe, Secure, and Ecologically Sound Storage Facility for Fissile Material Derived from the Destruction of Nuclear Weapons, October 1992, Washington, DC.
2. "Agreement Between the Government of the United States of America and the Government of the Russian Federation Concerning the Disposition of Highly Enriched Uranium Extracted from Nuclear Weapons," signed Feb. 18, 1993, Washington, DC.
3. Agreement Suspending the Antidumping Investigation on Uranium from the Russian Federation, Oct. 16, 1992.
4. Allied Signal and Nuclear Fuel Services representatives, personal communication, Nov. 16, 1992, including Allied Signal News Release, Allied-Signal Inc., Morristown, NJ, July 22, 1992.
5. Allison, Graham, et al. (ed.), *Cooperative Denuclearization: From Pledges to Deeds* (Cambridge, MA: Harvard University, January 1993).
6. Bukharin, Oleg, written comments to Peter Johnson, Office of Technology Assessment, Apr. 17, 1993.
7. Bunn, Matthew, Staff Director, National Academy of Sciences, personal communication, June 1993.
8. Burns, William F., Special Envoy on the Safety, Security and Dismantlement of Nuclear Weapons under the Soviet Nuclear Threat Reduction Act, testimony at hearings before the Senate Committee on Governmental Affairs, Mar. 9, 1993, and "Report to the Senate Committee on Foreign Relations on U.S. Efforts to Facilitate the Safe and Secure Dismantlement of Former Soviet Union Nuclear Weapons," Feb. 9, 1993. Progress report of the U.S. Delegation on Safe and Secure Dismantlement of Nuclear Weapons.
9. Campbell, K.M., A.B. Carter, S.E. Miller, and C.A. Zraket, "Soviet Nuclear Fission: Control of the Nuclear Arsenal in a Disintegrating Soviet Union," Center for Science and International Affairs, vol. 1, 1991.
10. Civiak, Robert, and Steve Goldberg, Office of Management and Budget, personal communication, Mar. 5, 1992.
11. Cochran, Thomas B., Senior Scientist, Natural Resources Defense Council, personal communication, June 1993.
12. Cochran, Thomas B., and Robert Standish Norris, "Russian/Soviet Nuclear Warhead Production," *Nuclear Weapons Databook* (New York: Natural Resources Defense Council, 1992).
13. Comptroller of the Department of Defense, Quarterly Report on Program Activities for Facilitation of Weapons Destruction and Nonproliferation in

the Former Soviet Union, covering the preceding quarter from January 1, 1993, through March 31, 1993, and cumulatively, Report to the Congress in accordance with section 108 of the FY 1992 Dire Emergency Supplemental Appropriations Act, Public Law No. 102-229, as amended.

14. Cook, Kevin, Assistant for Interagency and International Affairs, Department of the Army, personal communication, June 1993.
15. Davis, Zachary, and Jonathan Medalia, *Nuclear Proliferation from Russia: Option for Control* (Washington, DC: Congressional Research Service, Mar. 30, 1992), p. 17.
16. Factfile, *Arms Control Today*, vol. 22, No. 10, December 1992, p. 32.
17. Fialka, John J., ''Uranium from Russian Nuclear Arms Could Slash Energy Costs, U.S. Says,'' *Wall Street Journal*, July 17, 1992, p. A9.
18. Fulner, Robin, Defense Programs, and Roger Gagne, Production Analysis Program Manager, Nuclear Energy, Department of Energy, personal communication, Aug. 27, 1992.
19. Gagne, Roger, Office of Uranium Enrichment, U.S. Department of Energy, personal communication, May 27, 1993.
20. ''General Safety Criteria for the Russian Fissile Material Storage Facility,'' draft, produced at working meeting of Russian delegation, the U.S. Army Corps of Engineers, and Los Alamos National Laboratory, Dec. 11, 1992.
21. Glenn, Senator John, remarks at hearing before the Senate Committee on Governmental Affairs, ''Disposing of Plutonium in Russia,'' Mar. 9, 1993.
22. Goodby, Admiral James, Head of Safe, Secure Dismantlement interagency team, briefing to Office of Technology Assessment, Apr. 7, 1993.
23. Graham, Thomas, Acting Director, U.S. Arms Control and Disarmament Agency, comments at Nuclear Nonproliferation Workshop sponsored by the Congressional Research Service, Mar. 22, 1993.
24. Humphries, Marc, ''Restructuring the Uranium Enrichment Enterprise: A Comparison of S.2166 (Title X) and H.R. 776 (Title VI),'' CRS 92-271, Congressional Research Service, Mar. 6, 1992.
25. JARS-TEN-92-017, Sept. 21, 1992, ''Krasoyarsk Plutonium Plant Billed for Environmental Damage,'' PM0708151192 Moscow, *IZVESTIYA* (in Russian), Aug. 6, 1992, morning edition, p. 2.
26. Kay, David, Secretary General, Uranium Institute, former Chief Inspector for the U.N. Special Commission for Iraq under the International Atomic Energy Agency, comments at Nuclear Nonproliferation Workshop sponsored by the Congressional Research Service, Mar. 22, 1993.
27. Kelley, Joseph E., Director-in-Charge, International Affairs Issues, National Security and International Affairs Division, U.S. General Accounting Office, testimony at hearings before the Senate Committee on Governmental Affairs, Mar. 9, 1993.
28. Laughon, Kermit, Associate Deputy Assistant Secretary, Uranium Enrichment Program, U.S. Department of Energy, personal communication, Aug. 6, 1992.
29. Lepingwell, John, ''Beyond START: Ukrainian-Russian Negotiations,'' *RFE/RL Research Report*, vol. 2, No. 8, Feb. 19, 1993.
30. Lippman, Thomas, W., ''U.S. to Buy Russian Uranium,'' *Washington Post*, Sept. 1, 1992, p. A1.
31. Lippman, Thomas, and R. Jeffrey Smith, ''Russian Uranium Proposal Divides U.S. Government,'' *Washington Post*, July 23, 1992, p. A24.
32. Magwood, William, Edison Electric Institute, personal communication, Aug. 19, 1992.
33. Mah, Richard, Program Manager, Weapons Complex Reconfiguration Program, Los Alamos National Laboratory, comments at Nuclear Nonproliferation Workshop, sponsored by Congressional Research Service, Mar. 22, 1993.
34. McGoldrick, Fred, U.S. Department of State, personal communication, Mar. 26, 1993.
35. Natural Resources Defense Council, *Report on Third International Workshop on Verified Storage and Destruction of Nuclear Warheads*, Moscow and Kiev, Dec. 16-20, 1991 (Washington, DC: undated), 35 pp. plus appendixes.
36. Neff, Thomas L., consultant, personal communication, Feb. 28, 1993.
37. Norris, R.S., and William Arkin, ''Estimated Russian (CIS) Nuclear Stockpile (July 1993),'' *Bulletin of the Atomic Scientists*, July/August 1993, p. 57.
38. Nuclear Regulatory Commission, NUREG-1324, *Proposed Method for Regulating Major Materials*

Licensees (Washington, DC: Materials Regulatory Review Task Force, February 1992).

39. Nuclear Regulatory Commission, NUREG-1450, *Potential Criticality Accident at the General Electric Nuclear Fuel and Component Manufacturing Facility, May 29, 1991* (Washington, DC: Nuclear Regulatory Commission, August 1991).
40. NUKEM Market Report, Syposium Spotlight: World Experts Ponder Pathways to Warhead Dismantlement, highlighted presentations at the International Symposium on Conversion of Nuclear Warheads for Peaceful Purposes, jointly promoted by Science and Technology by the Development of Ethics and the A. De Gasperi Foundation, Rome, Italy, June 15-17, 1992, pp. 5-22.
41. O'Faircheallaigh, Ciaran, "Future Markets for Uranium—A Contribution to the Debate," *Communication in Energy Policy*, vol. 20, January 1992, pp. 74-77.
42. *Post Soviet Nuclear Complex Monitor*, "U.S. Assistance on Destruction of Soviet Weapons Moving Slowly," vol. 1, Feb. 24, 1993, p. 13.
43. Schmitt, Eugene, Director of the Office of Business Operations, Uranium Enrichment, Office of Nuclear Energy, U.S. Department of Energy, personal communication, Aug. 12, 1992.
44. Schutt, Paul F., Nuclear Fuel Services, personal communication, Jan. 12, 1993.
45. Sewell, Phil, Office of Nuclear Energy, U.S. Department of Energy, briefing to Office of Technology Assessment, Apr. 27, 1993.
46. Slipchenko, Victor, "Disposal of Fissile Materials Released as a Result of the Elimination of Nuclear Warheads," in Natural Resources Defense Council, *Report on Third International Workshop on Verified Storage and Destruction of Nuclear Warheads*, Moscow and Kiev, Dec. 16-20, 1991 (Washington, DC: undated).
47. Timbie, James, U.S. Department of State, personal communication, May 18, 1993.
48. Trout, John, Omaha District, Army Corps of Engineers, comments at Office of Technology Assessment Plutonium Storage Workshop, Feb. 10, 1993.
49. Trout, John, Omaha District, Army Corps of Engineers, "Comments on Issues for Discussion from the Office of Technology Assessment's Plutonium Storage Workshop," fax received Feb. 18, 1993.
50. U.S. Congress, General Accounting Office, *Uranium Enrichment, Unresolved Trade Issues Leave Uncertain Future for U.S. Uranium Industry*, GAO/RCED-92-194 (Gaithersburg, MD: U.S. General Accounting Office, June 1992).
51. U.S. Congress, General Accounting Office, *Russian Nuclear Weapons: U.S. Implementation of the Soviet Nuclear Threat Reduction Act of 1991*, GAO/T-NSIAD-92-47 (Gaithersburg, MD: U.S. General Accounting Office, July 27, 1992).
52. U.S. Congress, Senate, Report of the Committee on Foreign Relations, "The START Treaty," Executive Report #102-53, September 1992.
53. von Hippel, Frank, Professor of Public and International Affairs, Princeton University, personal communication, June 1993.
54. von Hippel, Frank, Frans Berkhout, and Harold Feiveson, *Report on the International Workshop on Nuclear Disarmament and Non-Proliferation: Issues for International Action*, Tokyo, cosponsored by Tokai University, the Federation of American Scientists, and the Program on Nuclear Policy Alternatives of Princeton University, Mar. 15 16, 1993.
55. *Weapons Complex Monitor*, "Russians Urge Collaboration on Disarmament Technology R&D," vol. 3, No. 21, Mar. 16, 1992.
56. White House Office of the Press Secretary, Fact Sheet on U.S.-Russian Agreement on Highly Enriched Uranium, Washington, DC, Aug. 31, 1992.

Policy Issues and Initiatives 7

In this post-Cold War era, retiring and dismantling warheads and managing warhead materials will constitute a new type of mission for the Federal Government. As discussed in chapters 2 and 3, the Department of Energy (DOE) continues to dismantle nuclear warheads retired by the Department of Defense (DOD) as it has in the past. Yet, important differences in the purpose of dismantlement—including the domestic and international context in which it is being conducted—render this a fundamentally different endeavor from what it was during the Cold War.

One of the major differences is that changed national security requirements and arms control agreements between the United States and Russia have diminished the need to maintain as large a stockpile of nuclear weapons. Thus, the present objective of weapons retirement and dismantlement both here and in Russia is to reduce the nuclear weapons stockpile, rather than—as in the past—merely to update and maintain it.

Further, there is now an international dimension to the weapons dismantlement and materials management effort. Decisions by the United States on these matters could either enhance or diminish the opportunity to reduce risks to international security and the environment. The United States is attempting to encourage Russia and other former Soviet Republics to dismantle their nuclear weapons and safely manage the materials coming out of warheads. At some point, the United States may need to point to its own accomplishments in this area to set an example for other nations to follow. In other words, success in this new purpose may be important not only in the domestic arena but also in the international one.

Point

"Here . . . is another example of the policy and political process going totally backwards—technical debate among experts, exploring limited sets of alternatives prior to the formulation of key government decisions about national goals and policies."

National public interest group reviewer of OTA report

Counterpoint

"There are always those who would like to make a grand task out of a rather straightforward one—politics is sure to make anything difficult."

DOE contractor official reviewer of OTA report

Failure to safely and securely dismantle retired warheads or to manage warhead materials here and abroad could have harmful consequences, in terms of both security and the environment, for the United States and the international community. To avoid these consequences, policies and programs are needed for dismantling warheads, and for managing the plutonium and highly enriched uranium from them, in a manner that reduces risks to national security, the environment, and human health.

The United States will have to define its national security objectives in the post-Cold War era with respect to nuclear weapons policy, and decide how best to conduct its own dismantlement programs so as to accomplish both domestic and international objectives. The United States will also have to develop the processes for determining what materials are surplus for weapons purposes and deciding how nuclear materials are to be managed over the long term. Finally, the United States will need to reshape its institutional structure so it can capably deal with nuclear materials from warheads.

At this time, the United States may still have an opportunity to assume a leadership role in ensuring that efforts to control the risks associated with nuclear weapons are initiated and carried out. However, if the United States does not act in a timely manner to define and implement a new policy of nuclear weapons dismantlement and materials management suitable to this post-Cold War era, changing international events may make it more difficult to have an impact on these matters in the future.

It is therefore important at this time that the policy basis for this new mission be declared at the top levels of Government, and that the mission be defined and articulated within the agencies responsible for implementing the policy. This can be done by the executive branch—initiated by the President and carried out by DOD and DOE—or Congress may take the initiative by directing the Administration to formulate a national dismantlement policy and providing guidance as to its scope and content.

Congress and the Administration can take a number of steps to help meet the technical, institutional, and political challenges of warhead dismantlement and materials management here, and to cooperate with Russia in meeting similar challenges there. The Office of Technology Assessment (OTA) has developed a series of policy initiatives that could improve prospects for successful nuclear weapons dismantlement and materials management programs. Although legislative action could appropriately be taken with respect to all the initiatives, and funds would have to be authorized and appropriated to carry out most of them, important steps toward accomplishing many of the initiatives can initially be taken by the President and the responsible executive agencies. The initiatives address the following matters that would be useful in improving the present situation:

- establish a *national dismantlement policy* that includes the objectives and scope of such dismantlement and a decision on the amount of materials from retired and dismantled warheads that will be declared surplus;
- *strengthen DOE management* of current and near-term activities to enhance worker and public safety as well as environmental protection functions;
- set up a process, and the appropriate institutional structures, for making and carrying out decisions about *the storage and ultimate disposition of nuclear materials from warheads*;
- establish a new organizational structure to manage long-term materials disposition and other activities;
- *increase access to information* regarding dismantlement and materials management decisions that are of public concern and interest; and

- determine approaches for *working cooperatively with Russia* to achieve mutually desirable outcomes.

These initiatives and some possible mechanisms for carrying them out, are discussed below.

A NATIONAL DISMANTLEMENT POLICY

The Nation needs, but does not have, a policy that sets forth clear, long-term goals for nuclear weapons retirement and dismantlement, and for management and disposition of materials from warheads. In addition to providing a broad framework for dismantlement and materials management, the Administration needs to decide on some specific features of future efforts. One such decision is the number of weapons to be retired and dismantled, and a time frame for completing dismantlement that is consistent with both safety and protection of human health and the environment. Another is the amount of plutonium and highly enriched uranium (HEU) currently available from dismantled warheads that is not needed to support nuclear weapons stockpile requirements, as well as the amount of such materials expected to accumulate incrementally and when dismantlement is completed. After these decisions are made, Congress and the Administration can begin to set objectives for long-term management of nuclear materials from warheads.

Openly developing and announcing a dismantlement policy could serve as confirmation to the international community that the United States is serious about disassembling most of its nuclear warheads and is willing to declare some of the materials from warheads as surplus no longer needed for strategic purposes. Statement of these policies will set an example that Russia's leaders could use to support similar dismantlement efforts there, as well as enhance prospects for further U.S.-Russian cooperation in setting and achieving mutual warhead dismantlement goals.

Setting Dismantlement Objectives

A dismantlement policy should specify the total number of weapons in the active stockpile that the Administration intends to retire and dismantle, and a time frame for the completion of their dismantlement. Announcing specific objectives and schedules will require careful coordination within the varied DOD and DOE organizational structures to make sure that all aspects are included and that the plan is practical, feasible, and safe from a technical and management standpoint. If a public document is issued, it will have to be carefully declassified, a task that will require additional time and resources.

The policy should clarify management responsibility and accountability within DOE and the contractor structure, and should provide uniform guidance to all field offices. Knowing what the entire dismantlement effort entails would allow for better planning and coordinating to incorporate safety and protection of the environment and human health into all operations. The projected cost of dismantlement could also be determined.

Determining and Disclosing Amounts of Surplus Materials

Once the number of warheads to be dismantled has been determined, the amounts of plutonium and HEU that must be managed will be known. An important part of the new dismantlement and materials management mission is to begin to move toward decisions about demilitarization and ultimate disposition of these nuclear materials from warheads. If policy is directed from the top levels of government, basic decisions about surplus materials could be made and announced in the near future, although the amounts may change as the international situation evolves.

Given the probability that the United States does not need to use all the plutonium and HEU from dismantled weapons to support stockpile requirements, it would be useful for DOE, DOD, and other involved agencies to determine as soon as possible the quantities of weapons plutonium and HEU that now exist—and are expected to exist after completion of warhead dismantlement—that will not be required for any future strategic uses. After a decision is made about what materials are surplus for such uses, those materials can be put on an irreversible path to demilitarization, and some preliminary actions can be taken that will lead to ultimate disposition decisions.

To encourage this decisionmaking process, Congress could direct that an unclassified report be submitted to Congress, and updated annually, specifying the amount of plutonium and HEU on hand from dismantled warheads that will not be needed to support stockpile requirements. An unclassified report on the materials allocated to nonstockpile use would serve to inform the public of the amount of material that must be dealt with in both the near and the distant future, and could facilitate analysis and public discussion of the safe storage and ultimate disposition of such materials. In this post-Cold War era, disclosure of the amounts of surplus materials should be possible without threatening national security or aiding potential proliferators. However, it would require careful study and a sound declassification process. Some initial information on possible future surplus nuclear materials has already been released by DOE and this policy could build on those first steps.

Designating materials as excess to stockpile uses would also lay the foundation for working cooperatively with Russia to encourage a similar policy of placing surplus materials from weapons dismantled there on a permanent demilitarization path.

STRENGTHENING DOE MANAGEMENT

For weapons dismantlement and materials management policies to be successfully implemented, DOE must carry out its responsibilities in a manner that satisfies the public that protection of the environment, health, and safety is being achieved. Success in dismantlement and materials management will be judged not only on whether DOE does its assigned job, but on whether it does it well from an environmental, health, and safety perspective. Thus, DOE must redefine its objectives in terms that will allow it to attain operational goals, while also meeting environmental requirements and expectations, and earning the public's trust. To do this, DOE must overcome the management assumption that, because dismantlement activities have been conducted for many years, nothing new or different is required in current or future operations. To ensure that DOE succeeds in doing so, both internal and external oversight of ongoing DOE activities will have to be strengthened.

Conduct of Dismantlement Programs

Although DOE is attempting to establish better environmental, safety, and health programs in its various operations, as described in chapter 3, the Defense Nuclear Facilities Safety Board has found that numerous problems persist where weapons dismantlement and materials management programs are being conducted. These findings point to a need for more vigorous efforts to upgrade environmental, health, and safety programs; to devote sufficient resources to these purposes; and to institute more effective training programs in connection with dismantlement and materials management. These factors strongly indicate that DOE; its Defense Programs Office, which has the responsibility for dismantlement; and its contractors who are involved in dismantlement and materials management programs will have to ensure that protection of the environment, safety, and health is a fundamental organizational objective.

To successfully accomplish this, there must be strong and visible management commitment to this objective at all levels—headquarters, DOE field offices, management and operations contractors, and subcontractors. The responsible DOE and contractor managers should institute a comprehensive environmental, health, and safety program in connection with dismantlement and materials management. The key elements of such a program are the establishment of specific policies and procedures, and their implementation in day-to-day operations; a credible and open internal evaluation process; and incentives for accomplishing the desired results. To succeed, the program must have adequate resources, including well-trained personnel, funding, facilities, and equipment. And, as discussed below, expanded external oversight is also needed.

DOE and its field offices should be more involved with planning and evaluating operations at the sites. A planning and evaluation process is needed within DOE that has purview over the entire dismantlement and materials management mission (including both Pantex and Y-12). DOE could establish an environment, safety, and health policy planning group that encompasses the total dismantlement and materials management program. That group could then issue clear guidance to managers as to how to implement the results of the planning process at all levels.

To achieve this new operational mode, workers must understand clearly how each program activity affects environmental and occupational safety and health objectives. Individuals with operational responsibility for actual dismantlement and materials management must take on "ownership" of the mission of protecting environment, health, and safety, and must be empowered to identify and seek solutions for problems as they arise.

To protect worker safety and health during dismantlement and materials storage operations, managers responsible for operations should assign occupational safety and health matters a high priority, implement effective worker protection strategies, enforce standards of the Occupational Safety and Health Administration (OSHA), and maintain clear chains of command.

Continuous efforts to institutionalize these changes are needed. Central to this is the development of operational practices and procedures with the participation of all relevant personnel, and continuous training of all personnel involved. It is also important to link personnel evaluations and rewards to environmental, health, and safety accomplishments, to provide strong incentives (monetary and otherwise) to achieve the desired results.

A vigorous internal audit and evaluation process is required to track accomplishments and identify areas for improvement. To this end, an organization and process should be established within DOE's Office of Defense Programs to conduct regular audits, and evaluate environmental, health, and safety awareness as well as results. In addition, DOE's Office of Environment, Safety, and Health should provide effective and continuous internal oversight.

External Oversight

To ensure continuous progress in the environmental, safety, and health aspects of current programs, it is necessary to have effective independent external evaluation and scrutiny by both technical experts and the public. Thus, it may be desirable for Congress to authorize additional oversight of ongoing dismantlement and materials management activities by the Defense Nuclear Facilities Safety Board, which already provides oversight of nuclear safety and related matters.

Congress may wish to give the Board responsibility to evaluate and approve Safety Analysis Reports and Operational Readiness Reviews for each weapons dismantlement program before it begins, and to monitor each program to ensure compliance. The Board itself could develop a process to ensure that environmental, health, and safety guidelines are being followed during weapons dismantlement and materials management activities. Areas that are not now emphasized—including environmental monitoring and environmental health issues—deserve particular attention. The Board would need additional resources and personnel to expand its activities in these areas.

Congress may also wish to direct the Board to provide more opportunity for public involvement in its oversight process. For example, instead of publishing final recommendations in the *Federal Register*, the Board could publish its draft recommendations and issue reports in unclassified versions, while more actively seeking public involvement in both processes. These actions would strengthen oversight and help assure communities that qualified outside parties are fairly and independently reviewing DOE's conduct of dismantlement and temporary storage activities.

With respect to worker health and safety, Congress could extend OSHA jurisdiction to DOE facilities. DOE's credibility would be enhanced if dismantlement and materials management activities were not subject exclusively to its jurisdiction. Until the necessary resources can be made available to OSHA, DOE's Office of Environment, Safety, and Health could invite it to conduct periodic reviews with respect to the warhead dismantlement and materials management program.

State oversight could be facilitated through better information exchange and increased funding. In addition, mechanisms through which interested citizens and experts could provide outside oversight at relevant sites should be considered. One such mechanism is establishment at a potentially affected site of a site-specific citizen advisory board, with full-time technical staff, to provide advice and input to DOE or other involved agencies on environmental, health, and safety issues relevant to warhead dismantlement and the temporary storage of warhead materials. This type of mechanism is discussed in greater detail in OTA's report *Complex Cleanup*, published in 1991.

NUCLEAR MATERIALS STORAGE

Permanent stockpile reduction means that fewer, if any, new weapons will be built in the future, which reduces the opportunity to recycle materials from disassembled warheads for use in other weapons. As a result, more materials from warheads must be managed for longer periods of time than in the past. In particular, additional storage of materials such as plutonium pits and HEU must be provided. Some amount of both plutonium and HEU will have to remain in "temporary storage," most likely for several decades.

At present, plutonium pits and HEU removed from disassembled U.S. warheads are in the custody of DOE's Defense Programs and are stored at its Pantex and Y-12 plants, respectively. DOE views storage of pits at Pantex as temporary (6 to 10 years) but has yet to announce plans or a timetable for any longer-term arrangements. In contrast, indications are that DOE intends to continue storing HEU from dismantled warheads at Y-12 in Oak Ridge for the indefinite future. As explained in chapter 4, regardless of what decisions are eventually made about the disposition of plutonium and HEU from warheads, storage of these materials for several decades will be required.

Planning for long-term storage involves resolving many technical, regulatory, institutional, and perhaps international issues—some of which are not within the purview and expertise of DOE. Yet, DOE's planning for long-term materials storage is not well coordinated with other agencies, and decisions being made in connection with the Weapons Complex reconfiguration could limit future options. As outlined in chapter 5, concerns have been expressed about prospective storage arrangements. Lack of public information on this subject has made it difficult for interested citizens to evaluate the adequacy of existing storage, as well as any plans for future modifications in current arrangements. Unless potentially controversial issues are identified and resolved early, it may be very difficult to implement long-term storage arrangements in a timely manner.

The schedule for Federal Government funding, design, and construction of a new facility, or even modifying an existing one, can take more than a decade. It would therefore be prudent to begin now to identify and evaluate alternatives that could provide safe, long-term storage arrangements for materials from warheads. Such alternatives might include upgraded arrangements or new facilities at various sites within or outside the Nuclear Weapons Complex. Once a national policy has been articulated, the next step is to determine how much plutonium and HEU will have to be stored over what period of time, and to provide capacity for that amount. Next, the form in which the material is to be stored should be determined since it is relevant to any future steps that would be taken. Finally, the infrastructure and materials processing capability necessary for a complete, safe, and modern storage facility must be provided.

Present plans appear to call for DOE to design a storage facility for plutonium pits that would be constructed on a Weapons Complex site and be operated by the Office of Defense Programs. Present plans also assume that the material to be stored in the new facility would be military in nature and thus not subject to licensing by the Nuclear Regulatory Commission or to regulation by OSHA. However, if many of the materials are at some point declared surplus, those assumptions would not necessarily apply.

Decisions about long-term storage of surplus plutonium and HEU from dismantled nuclear warheads involve many considerations that are not primarily within the purview of DOE. Such considerations include the scope of external regulation to which storage facilities and related activities should be subject, and the extent of transparency that may be needed to accommodate present or future U.S. foreign policy objectives. It would be desirable for these types of issues to be carefully considered early in the planning process by Federal agencies with primary responsibility for these matters. The early input of these and other agencies on relevant issues could help avoid problems later in the process.

To obtain this broad input, an interagency planning process could be constituted by the President to review the key issues surrounding storage of materials from dismantled U.S. warheads. One way to accomplish this is for the President to establish an interagency task force. If placed at a sufficiently high level within the Federal Government, this task force could bring a national focus to the important task of providing safe and secure storage of materials from dismantled weapons for as long as necessary. The task force could make recommendations regarding the most effective way to ensure safe and secure long-term storage of the materials from warheads. In addition, the task force could evaluate related issues such as the feasibility and consequences of storing the surplus plutonium and HEU separately from materials reserved for stockpile requirements, and the means for effectively involving the public in the siting of any new or modified facilities that may be needed in connection with long-term storage. To ensure that an interagency planning process takes place in a timely fashion, Congress could express its support for this process and direct the President to transmit recommendations about long-term storage of warhead materials to Congress by a specified date.

The task force could be composed of representatives from DOE (including not only the Office of Defense Programs, but also the Office of Environment, Safety, and Health, and the Office of Intelligence and National Security), DOD (including the Corps of Engineers), the Environmental Protection Agency (EPA), Nuclear Regulatory Commission (NRC), Department of Labor (particularly OSHA), State Department, Arms Control and Disarmament Agency (ACDA), and the National Security Council. Representatives of State regulatory agencies could be invited to participate as appropriate. The task force could be headed by an official in the Executive Office of the President and work for a period of approximately one year. Although this is an ambitious schedule, time is of the essence if the short-term storage problem is not to become by default a long-term storage crisis.

The task force should actively solicit expert and public views, possibly through hearings and interactive meetings with interested parties outside government, throughout the course of its work. Its draft reports could be submitted for public comment through the *Federal Register*. Extensive input from interested and potentially affected States should be solicited—possibly through an advisory panel composed of State representatives. Although these steps are time-consuming, they will help in obtaining relevant input early in the process so that issues raised can be properly addressed.

By expanding the planning process beyond DOE and including relevant expertise from other agencies, issues that DOE may not be factoring into its process are more likely to be considered. Environmental agencies such as EPA could encourage early consideration of issues such as pollution prevention, waste management, and environmental and radiation standards. Regulatory agencies such as the NRC should be able to bring to bear their experience in licensing analogous types of storage facilities. On the other hand, agencies that are actively involved in dealing with the international situation (such as the State Department and ACDA) could provide input about the likelihood and parameters of any transparency features that might be necessary. The States and the public could bring their perspectives to bear as well.

An important consideration in any future decisions involving storage is whether it would be feasible and useful to store surplus (nondefense) materials separately from those reserved for strategic purposes. A task force could weigh the advantages and disadvantages of separation from relevant viewpoints—international as well as domestic. The task force could also evaluate options for separate storage, such as whether such storage facilities should be located on the same or different sites, and whether strategic and surplus materials should be in the custody of the same or different agencies.

One potential consideration that should be weighed is whether separation of the two types of materials would facilitate transparency should bilateral verification or inspection be agreed on with respect to surplus weapons materials. In the case of plutonium pits, the task force could consider whether it would be desirable to "declassify" the shape and size of the pit, or the amount of plutonium in it, or whether the pits could be subject to nonintrusive monitoring or radiation measurements while in their containers. The task force could also consider whether physical separation of the civilian materials would more easily allow the United States to voluntarily render them subject to International Atomic Energy Agency (IAEA) safeguards—or perhaps even to place these materials under some level of IAEA management—should the United States and Russia ever reach an agreement to do so. Another consideration is whether separation of surplus materials from strategic ones would provide greater certainty to the public and to other nations that the surplus materials will not be used again for weapons.

Another factor for a task force to consider is whether, under current laws and regulations, the physical separation of U.S. civilian materials could facilitate subjecting these facilities and materials to the environmental, health, and safety requirements and standards applicable to all nonmilitary nuclear facilities—such as NRC licensing of any facilities in which the materials are to be stored—and also subjecting operations to NRC and OSHA regulation and oversight. The task force could also determine the legislative or regulatory changes that might be desirable to ensure that the materials will be stored and maintained under appropriate conditions so as to prevent accidents; protect the environment, health, and safety; and reduce worker exposure to radiation.

An important economic determination in terms of physical separation is the additional cost of security and safeguard measures. If civilian materials are stored at a different site from weapons materials, there will be duplication of security and safeguard services. Also consideration will have to be given to siting of a processing facility for plutonium pit maintenance, which DOE believes must be located at the same site as a long-term plutonium storage facility.

The interagency task force would also be in a good position to recommend a process for further involving the public in choosing options for modifying existing facilities or selecting locations for new, long-term storage facilities. As analogous problems (such as the Waste Isolation Pilot Plant in New Mexico and the high-level radioactive waste repository proposed for Nevada) have shown, arriving at a suitable and publicly acceptable process to select a site for facilities to contain toxic materials involves many institutional issues that can be even more difficult to resolve than technical matters. A carefully thought-out process is necessary to secure relevant input early and continuously. Any process for siting new facilities (both storage and associated processing facilities) or modifying existing ones should be developed and initiated now with continuous public involvement. The task force can bring to bear on this controversial matter the combined experience of the various agencies in public involvement.

Instead of delaying public discussion of longer-term storage until DOE's long-delayed reconfiguration process takes its course, or limiting public involvement to procedures associated with the National Environmental Policy Act process, it would be prudent to facilitate early public involvement so as to identify issues that need to be resolved. The interagency task force could solicit State and public input, consider the important factors from different points of view, define the issues to be resolved, and recommend a broad-based process for involving the public in discussions of the key issues before decisions are made and funds committed.

NUCLEAR MATERIALS DISPOSITION

Although DOE considers plutonium and HEU from weapons to be national assets, there is no national policy on what should be done with them. Options have hardly begun to be analyzed from a technical or institutional point of view, and there are wide differences of opinion about their merits, time and cost of development, and relative benefits. Today the discussion of disposition scenarios is often framed in terms of whether the materials are deemed "assets" or "waste." Rather than attempting to label the materials in this manner, it may be more useful to begin developing national policies and objectives for the disposition of surplus plutonium and HEU from warheads.

Thus, after a dismantlement policy has been established, a process should be initiated to bring together relevant governmental and nongovernmental views to provide the President and Congress with a comprehensive basis for making the policy decisions necessary before disposition of U.S. nuclear materials can be undertaken. Given the numerous political and technical uncertainties inherent in most of the approaches being discussed, definitive choices among potential options are difficult at this time. Nonetheless, it is prudent to start a process soon that looks toward long-term disposition of surplus U.S. nuclear materials, and considers the practical political and institutional realities of developing or applying particular technologies.

Although advocates of certain disposition technologies are optimistic that their favored technologies can be implemented successfully in certain time frames, many of the options being discussed at present are complicated and will be difficult to accomplish because economic and institutional issues, as well as technical ones, still have to be resolved. Institutional issues related to siting and other decisions have not been analyzed.

With respect to plutonium, storage will be required for several decades. Beyond that, however, it has not yet been determined whether plutonium should be stored indefinitely or used in some way (for either strategic purposes, commercial purposes, or both), or whether it should be disposed of expeditiously in whatever manner is feasible and acceptable. As indicated in chapter 4, some of the technologies being discussed in connection with plutonium disposition—such as advanced reactors and converters of various types—will take decades to develop, demonstrate, construct, and put into operation in the United States on a scale needed to accommodate the expected amounts of plutonium from warheads. Just as important, all of these technologies will produce waste in several forms—as spent fuel or other byproducts. At present, the Nation has no repository for such waste.

The situation with HEU, although somewhat different, also leads to the conclusion that storage will be necessary for a very long time. Although HEU can be blended down to low-enriched uranium (LEU) and used for commercial fuel, there is no present plan to do so with U.S. materials. Further, since the United States plans to purchase LEU blended down from Russia's HEU to fulfill some of its contracts for commercial reactor fuel, it is unlikely that the U.S. supply of HEU will be needed for this purpose any time soon. Therefore, storage will be necessary for these materials for some time to come, although eventually the United States may have to consider whether converting HEU for civilian use could have domestic or international advantages.

At present, as outlined in chapter 4, there are ongoing studies both within and outside DOE that address disposition scenarios for nuclear materials from warheads. Most of the discussions to date have been confined to the scientific community and have generally been conducted in technical terms. However, the analysis and selection of disposition scenarios encompass not only technical, but also public policy, issues. These issues range from facility siting, through the potential social, environmental, and economic costs and benefits of various approaches, to bilateral and international relationships, and perhaps ultimately even the future of nuclear energy. A process is needed for addressing these issues before decisions are made.

To get this process started, Congress or the President could create a national commission to recommend appropriate goals, policies, and programs relevant to the ultimate disposition of nuclear materials from warheads. A commission working over a period of approximately one year could gather the broad preliminary input from both public and private sources that is needed to inform the policy process. Such a commission should be composed of governmental and nongovernmental experts, as well as public interest and community representatives and other interested parties. This type of broad representation is needed to identify the wide range of issues involved in planning for ultimate disposition of nuclear materials from warheads. A commission would also provide an effective forum for identifying the relationship between potential disposition scenarios and U.S. policies and objectives here and abroad.

The commission could seek broad input on formulating both the objectives themselves and the plans that should be made to meet them. The objectives with regard to plutonium might well be different from those with regard to HEU—but in both cases, defining the objectives can help determine the most effective actions and the priority accorded them. The commission can also help determine the consistency of individual scenarios with these objectives and with broader national policies. This process could help avoid public expenditures on options that are inconsistent with national policies or objectives.

The commission could also outline the steps that might be needed to meet its recommended objectives. For example, an objective to convert plutonium as quickly as possible into a form that is less available or less usable for weapons would call for developing or applying technologies that can accomplish this in a reasonably short time frame. It may be useful in that case to select relatively near-term approaches using the most developed technologies such as vitrification, other modest processing, or available reactor systems.

If, however, the objective is to destroy as much plutonium as possible over the long term, then the plutonium would have to be retained in secure storage (and security costs would have to be considered) until more advanced technologies are developed that could bring about more extensive transformation of the elements or until other solutions can be implemented. In that case, the commission could attempt to define the level of effort needed to develop some of the advanced reactor or convertor options that would take longer to put into operation. Processes for handling and disposing of spent fuel or fission products would also have to be evaluated.

Advanced technologies associated with many of the disposition options will take decades to develop and cost a great deal of money. Before embarking on such programs, it is important to develop criteria against which options can be evaluated. Such criteria would include how well the processes and materials can be controlled and safeguarded, and how amenable they are to transparency in the interest of international cooperation. The commission could draw upon and expand the work of the interagency task force (discussed above) to develop and recommend criteria and indicate the priority that should be given to each.

Disposition of U.S. materials involves both short-term and long-term dimensions that require careful planning—particularly to prevent adverse environmental, health, and safety impacts. The commission could help identify and evaluate the relative advantages and disadvantages of individual options in protecting the environment, as well as human health and safety. It could also help assess technical availability—that is, how long, how expensive, and how difficult it would be to implement a particular option.

Most disposition scenarios would involve substantial processing of nuclear materials. Processing raises local environmental and public health concerns, as well as significant concerns about occupational health and safety. As discussed in chapter 5, one of the major objectives of public interest groups at DOE sites is to have the public included more actively, and earlier, in the decisionmaking process in order to better understand the health and environmental impacts of DOE activities on the community.

Meaningful involvement by an informed public and the States could help achieve acceptable outcomes. The commission could provide an early forum for the expression of community concerns and for preliminary discussion of the risks, as well as the advantages, of possible technical options. The commission could also recommend a process for facilitating public access to relevant information, and for early and effective public involvement in important decisions—including the location of facilities (for storage, processing, and disposal) that may be needed in connection with the ultimate disposition of nuclear materials.

Finally, the commission could examine the relationship of domestic to international scenarios. For example, any proposal to convert U.S. HEU to LEU and use it in commercial reactors would have to be evaluated in light of the LEU that the United States is planning to purchase from Russia for commercial reactor purposes. Taking these and other interrelated factors into account is necessary if disposition scenarios are to be evaluated properly.

The commission's work could provide useful input to future national policy decisions by Congress and the President. As discussed below, decisions will then have to be implemented by an organization with the capability to carry out the necessary activities in a manner appropriate to today's post-Cold War context.

A NEW MATERIALS MANAGEMENT ORGANIZATION

In addition to decisions on storage and disposition approaches, there is a need to develop and put in place an institutional structure capable of carrying out storage and ultimate disposition activities over the long term. Dealing with surplus plutonium and HEU from dismantled warheads over the long term is a mission that requires management to meet a new set of technical, institutional, and political challenges. In addition to possessing adequate technical skills, the organization that undertakes this mission should have the capability to carry out activities in a manner different from that traditionally associated with the defense mission at DOE.

For much of its past history, for example, the DOE weapons production mission, which is military in nature, was largely self-regulating. In contrast, the mission of surplus nuclear materials storage and disposition is largely civilian in nature and will not only have to be conducted from the start in compliance with environmental, health, and safety laws and regulations, but will also presumably be subject to appropriate outside regulation and oversight by Federal agencies such as the NRC, EPA, and OSHA, and by the States. Any facilities—including those for storage and processing—involved in materials disposition could presumably be licensed by the NRC and subject to EPA and OSHA regulation even if located within Federal facilities. Thus, it would be appropriate to have as the responsible organization one that can capably plan and carry out its activities in a regulated setting—a mode of operation that has not been the norm in the DOE weapons production organization.

Also, in the past, Weapons Complex operations have been conducted under extremely restrictive information access procedures. In contrast, the organization responsible for nuclear materials management and disposition—while continuing to protect properly classified information—will have to develop new approaches to information availability and access such as those described below. In a new organization, these approaches could be shaped on the basis of civilian, rather than military, objectives. In addition, a new organization could engage in a more open decisionmaking process to conduct credible environmental and related analyses in connection with the development and selection of technologies, the selection of sites, and the management of ultimate disposition activities.

Further, the new mission of materials storage and disposition may have to be conducted in a manner consistent with whatever bilateral or international considerations become relevant in the future. A new organization dealing with ultimate disposition of nuclear materials from warheads could start with a well-defined mission that includes not only domestic but also international imperatives, and be structured from the outset to integrate international considerations into its programs and actions.

The responsibility for materials disposition could be assigned to a new organization that is given both the clear and primary mission of developing and implementing plans for ultimate disposition of surplus nuclear materials from warheads, and the personnel qualified to carry out that mission. Congress may wish to establish a new organization for materials management and disposition as a separate agency outside DOE. Or a new office could be created within DOE—separate from and independent of the Office of Defense Programs—with a strictly civilian, nonweapons mission.

Alternatively, such an organization could be established within another Federal agency. To avoid past practices that led to widespread contamination at nuclear weapons sites, the new organization must take appropriate measures to conduct activities so as to minimize health and environmental risks. Because past polices have led to the inability to locate or operate nuclear waste repositories, the new organization will have to gain the acceptance and support of the American public. It is therefore important to develop a decisionmaking process that is open, fair, and responsive to public concerns, and the managers of the organization should be selected in part on the basis of their ability to operate in a mode that stresses openness and public involvement.

Although the establishment of a new materials management and disposition organization will take some time, the process need not delay ongoing dismantlement activities. The transition to a new organization could begin to be planned at any time, and its implementation could await the formulation of a national dismantlement policy and completion of the interagency planning process for long-term materials storage discussed above.

An important advantage of giving materials disposition responsibilities to a new organization is that it can start with a "clean slate" and consciously approach its mission in a manner that would earn the public's trust and confidence. A new organization would have the advantage of being able to give priority to institutional matters such as interaction with the public, building credibility through early and effective public involvement, and being genuinely responsive to public concerns. A significant advantage of a new organization that has a purely civilian mission is that it can deal with materials disposition issues pertinent to U.S. relations with Russia and the rest of the international community on civilian rather than on military terms.

INFORMATION ACCESS

As discussed in chapter 6, the present institutional context for dismantlement and materials management differs from that of the past. Although DOE still enjoys some degree of self-regulation over its nuclear activities under the Atomic Energy Act, many of its activities are now subject to environmental regulation and safety oversight. There is also more public scrutiny of whether DOE conducts these operations safely, with maximum protection of human health and the environment. To assure the public that this is being done, DOE (or a new organization discussed above) will have to make more information available than has been done during past activities dealing with nuclear weapons production.

In the interest of national security, legislative requirements have long prohibited public access to a broad range of information related to nuclear weapons. DOE and DOD also have discretion to further limit information access related to these and other matters. While restricting access to data on nuclear weapons design and manufacture that could aid proliferators and terrorists continues to be important, the end of the Cold War raises the question of whether current restrictions on access to information that may be relevant to dismantlement and materials management are still necessary.

This question is particularly relevant when it comes to disclosure of information regarding environmental, safety, and health issues associated with nuclear weapons dismantlement and materials management and disposition. Data that citizens consider essential to discussions of environment, safety, and health are often not accessible to interested persons outside DOE, because the data may be contained in documents that also contain classified or otherwise controlled information. A lengthy and meticulous review process is necessary to remove even small amounts of classified information from documents in order to release unclassified information to the public. As a result, the ability of the public to acquire adequate and timely information related to these activities has been impaired.

Another problem facing citizens is the generally slow and often inadequate responsiveness of DOE to many legitimate requests for information. Citizens frustrated by lack of information access are not likely to trust the agency in question or support its plans and programs. Yet such trust and support are critical if warhead dismantlement and materials disposition programs are to gain needed public acceptance.

Although there are several ongoing efforts within the executive branch—and particularly within DOE—to review information classification procedures and increase public access to information, Congress may wish to consider whether the existing legal basis for restricting access to information is appropriate in light of today's post-Cold War national security objectives. Congress may also wish to provide additional resources to accelerate the declassification of documents and the enhancement of public access to information. In particular, Congress could consider enacting updated standards for dealing with information relevant to nuclear warhead dismantlement and materials management or disposition—particularly data relating to environmental, health, and safety matters—and require agencies to promulgate rules and adopt procedures consistent with those standards.

In preparation for possible legislative revisions, Congress could request DOE, DOD, and other involved agencies to prepare joint recommendations for comprehensive legislative provisions regarding access to information relevant to nuclear weapons dismantlement and materials management and disposition, particularly with respect to environmental, safety, and health issues. This effort could include a review of Atomic Energy Act provisions, and implementing rules and regulations dealing with "Restricted Data," "Formerly Restricted Data," and "Unclassified, Controlled Nuclear Information" (UCNI). The agencies could be requested to recommend whether repeal or modification is necessary or appropriate in light of changed national security requirements. Recommendations could also be made for new legislative standards that would constitute the exclusive basis for restricting access to data and documents and for disclosing or disseminating information.

The review could also encompass the standards and procedures for national security classification and other categories of restrictions that may apply to nuclear warhead dismantlement and materials disposition. Depending on the nature of the recommendations, they could be implemented by revising existing laws, enacting a new law that would supersede conflicting sections of existing laws, issuing an Executive order, or promulgating new rules and regulations pursuant to the notice and comment requirements of the Administrative Procedures Act.

The rules could, for example, set up procedures for allowing general declassification requests for certain types of information to be made by parties outside DOE. The rules could also establish procedures that place the burden on those who wish to retain classification status or other restrictions on information, and provide for resolution of issues in favor of disclosure. Since this may entail a more difficult and complicated process than exists at present, enhanced resources will be required.

Pending consideration and enactment of needed fundamental changes, Congress could require DOE to accelerate efforts to declassify or remove from the UCNI category appropriate documents that contain information relevant to issues of environment, safety, and health. Also, to emphasize the importance of information access, there could be established within DOE's Office of Defense Programs an "office of declassification" that would be responsible for issuing unclassified versions of classified documents and unrestricted-access versions of UCNI information.

There could also be established—within DOE's present Office of Classification—an "office of information access." The new office could be given the task of promptly investigating and addressing any complaints from the public about the timeliness of information requests regarding weapons dismantlement and materials disposition issues under the Freedom of Information Act or other procedures. The office could also anticipate public interest in forthcoming reports about dismantlement and materials disposition and related health and safety matters, and make sure that unclassified versions are issued.

Increasing information access at the agencies involved in warhead dismantlement and materials management would give interested parties the ability to examine the relevant data and to intelligently evaluate proposed decisions about warhead dismantlement and materials management and disposition. This in turn could result in more informed and effective public input to DOE's decisionmaking process.

COOPERATION WITH RUSSIA

The United States cannot single-handedly reduce or eliminate all the security and environmental risks posed by nuclear weapons and materials since it does not control all weapons and materials worldwide. At present, most of the non-U.S. weapons and materials are in Russia.[1] As described in chapter 6, the United States has initiated a program of assistance to Russia with regard to safe, secure dismantlement, including the construction of a plutonium and HEU storage facility. In addition, an agreement is pending whereby Russia will convert its HEU from weapons into LEU and sell it to the United States. No clear policy exists in Russia for ultimate disposition of plutonium from weapons, but reports indicate that many officials favor technologies that will use the material in power-producing reactors.

The United States at present has not been able to verify what weapons have been or are planned to be dismantled in Russia, what quantities of weapons-grade nuclear materials exist or are being produced, and what is being done with materials from dismantled warheads. The United States will presumably maintain appropriate security and accountability systems for its own nuclear weapons and materials. However, similar results may not be easy to achieve abroad. Given these circumstances, the United States will have to decide whether its present initiatives and programs to assist Russia are adequately promoting national security objectives, or whether there is a need to develop different or additional approaches.

Past U.S. efforts to assist Russia with dismantlement and materials management have provided important opportunities for cooperation, but there is not yet a mature process for coordinating and focusing policy at the highest levels of government. The United States and Russia should agree on the most important immediate and long-term objectives of each with respect to warhead dismantlement, and materials management and disposition. It is important to develop a plan for attempting to reconcile differing national objectives and requirements. Without such a plan, the United States may be implementing programs to assist Russia that have conflicting goals, lead to unexpected or undesired consequences, or do not address the most important issues.

Russia has indicated that it does not want the United States directly involved in the dismantlement of its warheads and that it wishes to construct its own materials storage facility (with U.S. design assistance). On the other hand, Russia has been willing to enter into an agreement with the United States on plans to sell uranium from Russian warheads and has not ruled out the possibility of cooperative efforts on approaches to plutonium disposition. There is now an opportunity to pursue such efforts, but it is not clear how long that opportunity will exist.

Developing a Strategy for Cooperation

It is in the national interest to seize the present opportunity and work cooperatively with Russia to achieve desirable outcomes, particularly with respect to plutonium disposition. A process is needed to ascertain the feasibility of U.S.-Russian cooperation in materials disposition efforts. If more active U.S. involvement were deemed prudent, choices would have to be made regarding the technologies that need to be developed or applied to meet desired objectives, and the amount and sources of funding to be devoted to these purposes would have to be determined. There may be fiscal benefits to establishing such policies early, and technology development could be focused on helping to achieve disposition options for Russian nuclear weapons materials consistent with international security objectives.

Since any cooperative programs involving Russia and the United States will ultimately

---

[1] While this report does not specifically address nuclear weapons in Ukraine, which are the subject of diplomatic efforts, U.S. initiatives similar to those discussed here could be considered with respect to Ukraine.

depend on mutual trust between the two nations, it is important for each nation to have assurance that the storage and disposition of nuclear materials from dismantled warheads are being managed in a manner that is safe and secure, and that protects against proliferation and prevents future use of the materials in weapons. Some type of reciprocal information disclosure and verification could greatly assist in strengthening and maintaining mutual trust and cooperation.

U.S. initiatives to assist Russia in materials management and disposition efforts so as to prevent proliferation are more likely to succeed if they are guided by a high-level, focused, national policy that calls for the United States to work cooperatively with Russia to achieve mutually acceptable disposition of both nations' surplus materials. To strengthen prospects for United States and Russian cooperation with respect to the disposition of materials from warheads, the President and relevant Executive agencies need to increase coordination at the highest levels to develop a strategy and plan for this purpose. The strategy should be aimed at identifying key issues and reconciling differing national objectives, requirements, and priorities. This would serve as a foundation for developing a mutually acceptable plan for disposition of Russian plutonium and perhaps for U.S. assistance in that effort.

Key components of an effective strategy should include strengthening the relationship between U.S. assistance in materials disposition programs and other programs in which assistance is desired by Russia; and developing and negotiating with Russia an initiative for mutual disclosure of information and for reciprocal arrangements to verify the amounts and monitor the status of these materials over the next decade. Another element of a strategy could be to strengthen the link between U.S. and Russian efforts in nuclear materials management and disposition.

Strengthening Cooperative Efforts

One approach to cooperating with Russia on key issues is for the United States to help Russia understand the nature of U.S. materials management and disposition efforts. Although U.S. objectives with respect to its own weapons plutonium and HEU may be quite different from Russia's, joint deliberations on these subjects and joint technical studies may help Russia to understand what the United States is doing and to evaluate its own options. To accomplish this, coordination is especially necessary between agencies responsible for U.S. materials management and disposition programs and those responsible for U.S. policy toward Russia, and initial steps should be taken toward conducting joint studies and negotiating some reciprocal arrangements for materials monitoring.

To develop an effective relationship between U.S. and Russian programs, agencies (such as DOE) that are knowledgeable about U.S. materials management and disposition will have to work more formally and continuously with agencies (such as the National Security Council and the State Department) that shape and conduct U.S. relationships with Russia and other former Soviet republics on these matters. Strengthened intergovernmental coordination would help ensure that programs for U.S. dismantlement and materials disposition—as well as programs to assist and work with Russia in these matters—are carried out consistently with U.S. policies and are aimed at achieving mutually acceptable goals.

A joint U.S.-Russian effort to analyze and evaluate nuclear materials disposition technologies is one approach that could provide useful insights into the important immediate and long-term objectives of the United States and Russia with respect to the disposition of weapons materials. Such a project could be conducted over a 2-year period and be directed jointly by individuals from the United States and Russia, and the project team could include persons from all the relevant disciplines and also draw upon expertise in Europe and in other nations. The team's work could be reviewed by an advisory panel composed of members from the Government, national laboratories, and industry; the scientific, academic, and environmental communities; and the public sector in the United States, Russia, and Europe.

The project team could also consider formulating a proposal for U.S. participation in developing and applying technologies deemed necessary to implement feasible disposition options in Russia, particularly options that could help demilitarize plutonium as soon as possible. In addition, technology sharing among the United States, Russia, and European or other nations with regard to disposition options can also help achieve cooperation in the ultimate disposition process, perhaps avoid options that could lead to proliferation, and help settle on approaches that promote nonproliferation as well as protection of the environment and health. It could also provide a foundation for future cooperative participation by the U.S. government and the private sector. A clear U.S. objective to encourage and assist Russia in converting plutonium, as soon as possible, into a form not usable for weapons could reasonably involve U.S. participation in developing technologies for that purpose—regardless of whether those technologies would be used for the disposition of U.S. materials.

Another approach may be to link efforts to work cooperatively on nuclear materials management and disposition with other efforts to assist Russia in some of its critical needs and interests not related to nuclear weapons dismantlement. For example, there are pressing environmental restoration needs in various parts of Russia. More active U.S. assistance in this and other areas deemed important by Russia could be part of the overall assistance offered with respect to materials management and disposition. Also, since political instability and economic concerns could directly affect the success of Russian dismantlement and materials disposition efforts, the United States may decide that spending more funds in Russia would actually decrease the risk from nuclear weapons if it resulted in increased economic and political stability there. A flexible approach that aims to assist with some of Russia's priority needs may be more successful than some of the present programs, and could lessen criticism in Russia that U.S. programs are diminishing its international status or that the United States is attempting to dictate specific aspects of Russia's nuclear enterprise.

To strengthen and maintain mutual trust and cooperation, the United States should develop means, consistent with recent legislative provisions, that would enable Russia and the United States to exchange information about nuclear materials from warheads. This initiative could apply to the plutonium from dismantled warheads and also supplement any transparency arrangements made in connection with the agreement for the sale of Russian LEU to the United States for use in the commercial power market. Exchanging information about amounts of nuclear materials in the custody of each nation would help the United States better understand the dimensions of Russia's problems in managing these materials. Information exchange would enable the United States to offer any relevant assistance in accounting for the materials or keeping them safe and secure through whatever disposition processes Russia undertakes. Presumably, such information will not be volunteered by Russia without some offer of reciprocity on the part of the United States; thus a U.S. initiative in this direction will be needed.

In addition, Russia may soon begin to look to disposition options for its plutonium—options that may involve moving this material from place to place, processing it, and changing its form. U.S. concerns about preventing any of the Russian materials from getting into the wrong hands or causing serious environmental harm could best be addressed through arrangements that permit the United States to verify the amount and status of the plutonium in these various processes and to monitor the processes as appropriate. Again, it is unlikely that Russia will agree to such monitoring without some reciprocity with respect to U.S. materials.

Thus, if an arrangement can be worked out that will not compromise the national security of the United States, bilateral reciprocal agreements entered into as soon as possible could forestall potential problems with respect to the nuclear materials from warheads in the former Soviet Union. Such arrangements would also enable the United States and Russia to cooperate in the common goal of preventing the proliferation of nuclear weapons. Finally, this approach could be effective in pursuing U.S. goals of international security.

CONCLUSION

The policy initiatives discussed in this chapter were developed from OTA's analysis of the problems facing current efforts in dismantlement and materials management, as well as the opportunities that exist today to move forward and make significant progress in stockpile reduction and control of future risks. Box 7-A summarizes these conclusions and initiatives. OTA's analysis indicates that each initiative has individual merit and could be implemented separately. Alternatively, several or all of the initiatives could be implemented collectively. Congressional action in the form of legislation or oversight could begin the process for each, but most could also be implemented by the Administration directly. Success in safe dismantlement and adequate protection of people and the environment for generations to come is an important national goal—the Federal Government has a serious responsibility to meet that goal.

Box 7-A—Key Policy Conclusions and Initiatives

1. A National Dismantlement Policy

Conclusion

Although the responsible Federal agencies (DOD and DOE) have been carrying out warhead dismantlement and view this as business as usual, dismantlement and nuclear materials management in today's post-Cold War context constitute a new type of mission for which a national policy is needed to guide future actions.

Initiative

Develop and announce a national policy that sets goals for warhead dismantlement and materials management, and specifies the amount of plutonium and highly enriched uranium (HEU) from dismantled warheads that will not be needed to support future stockpile requirements.

2. Strengthening DOE Management

Conclusion

Although DOE is attempting to make improvements in its environmental, health, and safety practices, more attention still needs to be devoted to these matters in connection with warhead dismantlement and materials management.

Initiative

Implement a DOE management system that gives priority to protecting the environment, health, and safety; expand and strengthen external oversight of DOE dismantlement and materials management activities by independent outside entities.

3. Nuclear Materials Storage

Conclusion

Since storage of surplus nuclear materials from warheads will be needed for several decades, and many issues outside the purview of DOE will have to be addressed before long-term storage can be implemented, an interagency planning process to identify and resolve these issues should begin as soon as possible.

Initiative

Establish an interagency task force that includes Federal agencies with expertise in regulatory, international, and public involvement matters to recommend a plan for safe and secure storage of nuclear materials, and to develop a process acceptable to the interested public for siting new or modified storage facilities.

4. Nuclear Materials Disposition

Conclusion

Consensus is lacking about whether surplus warhead materials should be stored indefinitely, converted into forms usable for commercial power generation, or disposed of as waste, and about the technical, economic, and political merits of various disposition options and technologies, and a process for openly discussing and reconciling diverse governmental and nongovernmental perspectives on these issues is needed.

Initiative

Create a national commission to recommend goals, policies, and programs for ultimate disposition of surplus plutonium and HEU from warheads, and to provide a basis for developing an ultimate disposition policy for these materials.

5. A New Materials Management Organization

Conclusion

Since carrying out programs for the ultimate disposition of surplus plutonium and HEU from dismantled warheads is essentially a civilian mission that will require not only technical skills, but also the ability to meet a new set of institutional and political challenges, the DOE organization that has been carrying out the weapons production mission is not well suited to take on the new materials disposition mission.

Initiative

Create a new organization outside DOE to manage surplus materials from warheads, or establish a new organization for this purpose within DOE or some other existing agency.

6. Information Access

Conclusion

Some of the restrictions on access to information relevant to warhead dismantlement and materials management may no longer be required in this post Cold War context, and enhancing information access could increase public trust and confidence with respect to these activities.

Initiative

Review and possibly revise the existing legal basis for restricting access to information in light of today's post-Cold War national security objectives, and accelerate efforts to increase access to information relevant to warhead dismantlement and materials disposition.

7. Cooperation with Russia

Conclusion

Although important steps have been taken to assist Russia with weapons dismantlement and materials management, a focused and coordinated strategy within the Federal Government can improve prospects for cooperating with Russia to develop a mutually acceptable plan for disposition of its plutonium.

Initiative

Strengthen the relationship between U.S. assistance to Russia for materials disposition and other programs in which assistance is desired by Russia; develop a means for joint assessment of plutonium disposition technologies; and negotiate mutual disclosure of information and reciprocal materials monitoring arrangements.

Appendix A: Plutonium Pit Storage at Pantex

For more than four decades, plutonium pits from nuclear weapons disassembled at Pantex were stored in concrete bunkers[1] prior to shipment to the Rocky Flats Plant in Denver. The same concrete bunkers were also used for briefly storing pits fabricated or recycled at Rocky Flats (30). After weapons disassembly, plutonium pits are placed in drums and transported to bunkers for storage in rows 3 drums wide and 23 drums deep on one side of the bunker, and 2 drums wide and 23 drums deep on the opposite side (19). A 4-foot aisle separates the rows of drums (24). The brief storage of plutonium pits at Pantex is commonly referred to by Department of Energy (DOE) officials as "staging."

Recent reductions in the weapons stockpile and in weapons component production, downsizing of the Nuclear Weapons Complex, closing of plutonium processing facilities at Rocky Flats for environmental and safety reasons, and the little additional storage space available at other facilities, all have contributed to a shortage of pit storage space throughout the Nuclear Weapons Complex (7). This in turn has placed additional pressure on DOE to store plutonium pits at the Pantex Plant.

DOE predicts that once the storage capacity provided by the "most preferable configurations" in use has been reached, all weapons still in the custody of the Department of Defense and planned for disassembly at Pantex will have to remain stored at military installations. Weapons delivered to Pantex that were not disassembled because of pit storage limitations will remain in the particular bunkers where they are staged. This storage limit may be reached in late 1993, and even if packing density is increased, the limit will be reached in mid-1994 (24).

EFFORTS TO INCREASE STORAGE CAPACITY

Consideration of the National Environmental Policy Act Process

To increase the pit storage capacity at Pantex and avoid delays in the dismantlement process, DOE plans to increase the storage capacity of the Pantex bunkers. As part of the National Environmental Policy Act (NEPA) process to institute this change, DOE has prepared a draft Environmental Assessment (EA) under the act. This EA evaluates any environmental impact that might result from increasing both the number of bunkers that can be used for pit storage and the total number of pits that can be stored per bunker. To achieve this objective, DOE had planned to issue

[1] Originally designed to protect conventional ammunitions from bomb blasts during World War II, bunkers are two-room storage facilities made of concrete, covered with earth—except for the front door—with a floor capacity of approximately 1,040 square feet. These concrete bunkers are technically known as Modified Richmond. A double-leaf steel door with two sets of security locks is used in each room to prevent unauthorized entrance. Portable radiation monitors must be carried by personnel when entering bunkers.

Because the number of concrete bunkers now dedicated to plutonium pit staging is insufficient for the increasing number of pits that will result from planned weapons dismantlement operations, the Department of Energy has proposed using some of Plant 42's steel arch construction storage facilities—also known as SAC magazines—that stage nuclear weapons or sensitive weapons components (26).

for public comment, by August 1992, a draft Environmental Assessment evaluating plutonium reconfiguration. Because of delays, the draft EA was not issued until December 1992, and as of this writing, DOE has received and is evaluating numerous public comments on that draft.

Earlier in 1992, DOE officials indicated that if the EA identified any significant environmental impact associated with the proposed expansion of plutonium pit storage at Pantex, a more detailed Environmental Impact Statement would be prepared. According to the same officials, implementation of multiple stacking configurations for plutonium pit storage will not begin until the proper environmental and safety documentation on plutonium storage has been completed (15). If no potential impacts are identified, DOE will implement its proposed expanded storage configuration for plutonium pits following the issuance of a Finding of No Significant Impact (19).

Preparation of a Safety Analysis Report

To evaluate the safety of bunkers proposed for expanding plutonium pit storage, DOE prepared a Safety Analysis Report (SAR) in 1992. As part of the study, DOE examined the structural safety of bunkers, the potential causes of accidents, and their consequences to plant workers, the environment, and the general public. Some of the accident scenarios evaluated in the report include those initiated by earthquake, human error, aircraft collisions, tornadoes, and fires. Access to this document for public review has been restricted by DOE's information classification rules.[2]

Improvements in Storage

DOE is funding several projects to increase its understanding of the handling, safety, and design requirements of plutonium pit storage. For example, a computer model was developed by Sandia National Laboratory to identify ideal packaging configurations and optimal space use for storing pits in bunkers (8).

Pantex is also working with the national laboratories to improve the safety of plutonium pit storage at bunkers and reduce radiation exposure of personnel during loading, unloading, handling, and inventorying operations. One such effort is the project known as "Stage Right," in which several Sandia National Laboratory scientists are studying the potential application of robotics in activities such as handling and monitoring pits inside the bunkers. First, the project is to assist the development of a manually operated, shielded forklift to handle pallets of pit-containing drums and thus reduce radiation exposure (13). It is estimated that radiation exposure levels could be reduced by a factor of five (3,24). Completion of the first forklift development project is expected in 1993 (14).

The next step in the forklift project would be to design and manufacture an unmanned forklift to remotely control the loading and unloading of pit containers in bunkers. Eventually, an automatically guided vehicle could be developed, capable of handling and monitoring containers, thereby reducing radiation exposures even further (3,13).

Sandia National Laboratory is responsible for two additional projects relevant to plutonium pit storage at Pantex. One involves the design, fabrication, and demonstration of the system to be used by the forklift during multiple stacking of plutonium pits. The other project concerns the development of a system for performing plutonium accountability activities, such as inspections, without having to enter the bunkers. All of these development programs, however, must await a decision by Pantex management before being incorporated into actual operations.

CHANGES EXPECTED IN STORAGE

Pantex has disassembled more than 50,000 weapons in the last four decades (15). Although the manner in which disassembly activities are conducted has not changed, storage practices and facilities at Pantex will have to change. Some of the changes likely to result from increased plutonium pit storage include the following:

- Under the planned increase in weapons disassembly rates, the number of bunkers employed for pit storage could more than double, while the number of pits stored could reach a total of 20,000 (24). The bunkers to be added for this purpose (known

[2] The SAR was determined to contain Unclassified Controlled Nuclear Information (UCNI). DOE orders restrict the distribution of UCNI documents mainly to Government officials or certain other official groups who have been designated to have a need to know and will also agree to control the document.

as steel arch construction or SAC magazines) have been used to store weapons and weapons components.

- In its Environmental Assessment, DOE proposed to increase storage capacity using a "horizontal palletized multiple configuration" to store up to 440 pits per bunker (24). Figures A-1 and A-2 show, respectively, the single and multiple stacking configurations associated with plutonium pit storage at Pantex.
- The increased emphasis on weapons disassembly and pit storage at Pantex may result in operational changes at the plant; however, little information exists regarding the type and extent of such changes. According to a draft internal policy letter, the plant operating contractor (Mason & Hanger) expects radiation exposure for a worker involved in weapons dismantlement operations "... to be significantly greater than the collective dose that radiation worker received in 1991 [primarily as a result of] dismantling more weapons from programs with higher dose rates." According to this document, Mason & Hanger plans to institute measures to ensure that individual exposures do not reach more than 20 percent (or 1 rem) of the maximum allowable DOE radiation dose limit (1).
- Total radiation exposures of workers responsible for conducting periodic inspections and inventories of drummed pits could increase. If interim storage were to involve all 60 bunkers with a single-layer vertical configuration, storage of up to 20,000 pits would result in an annual collective dose of 100 to 200 person-rem—an increase of four to eight times the levels measured in 1987-91. However, DOE calculates that the use of a palletized multiple stacking configuration will result in a collective dose of 50 to 100 person-rem per year, or only two to four times the 1987-91 levels.
- If no changes occur in current inspection practices, workers could also experience increased radiation exposure levels if: 1) the protective equipment (lead apron) worn does not protect a worker's back and extremities; 2) radiation exposure is not monitored and reported in terms of real time but rather on a monthly basis; or 3) workers experience difficulty in identifying specific pit-containing drums and thus increase work time (3). Plans are under way to reduce these occupational radiation exposures through the use of shielded forklifts, bar codes and bar code readers, and eventually, unmanned vehicles and robots (24).
- Available information about pit storage configurations used in the past is unclear. The information provided to public officials and communities about the configurations used for pit storage at bunkers has led many to believe that the maximum number of pits that could be safely stored under the single-layer, vertical position configuration was 240 per bunker. DOE, however, has used other configurations and has stored more

Figure A-1—Pit-Filled Concrete Bunker With Single-Layer Storage Configuration

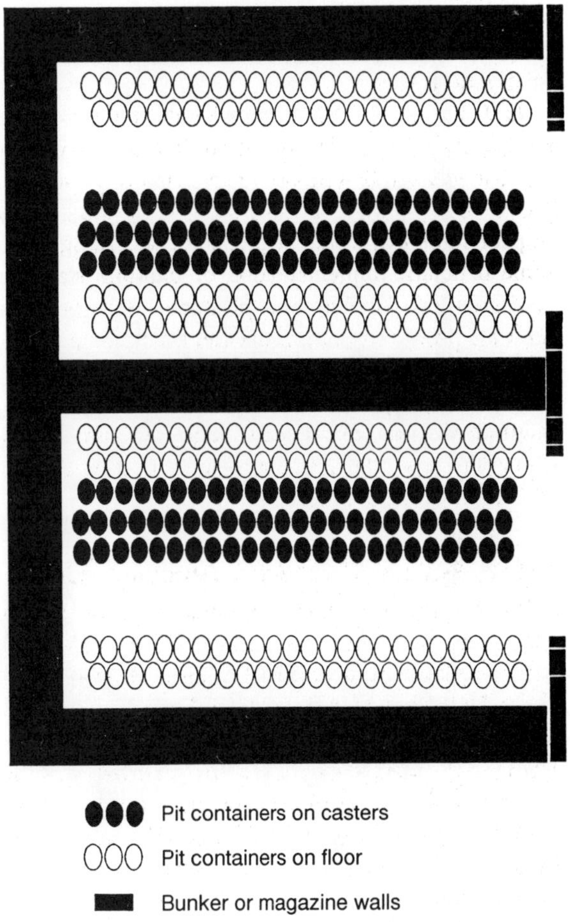

SOURCE: U.S. Department of Energy.

Figure A-2—Multiple Stacking Configuration Proposed for Plutonium Storage at Concrete Bunkers

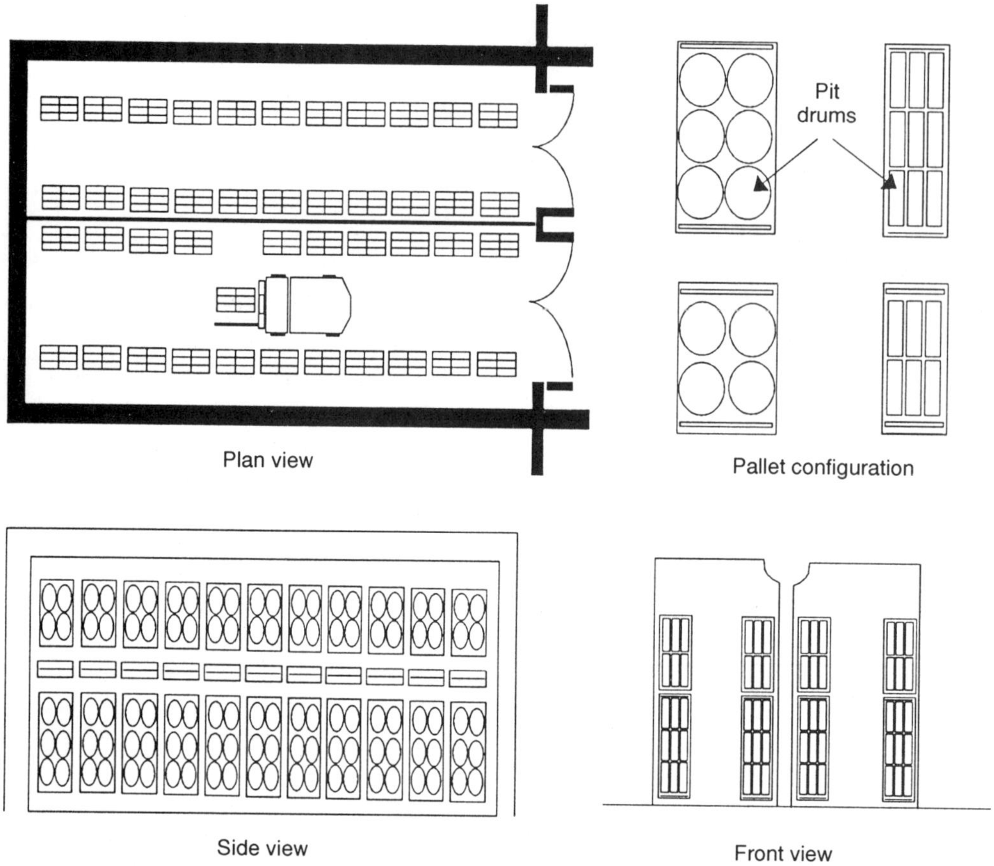

SOURCE: U.S. Department of Energy.

pits by putting additional drums in the space available between the rows of stored drums (19,25). This has led to public concerns about what other configurations DOE might use to extend available pit storage capacity at operating bunkers.

PUBLIC CONCERNS

In December 1992, DOE submitted a predecisional draft of the Environmental Assessment for interim storage of plutonium components at Pantex to the State of Texas for review and comments. The Texas Governor's office, in turn, distributed the EA to a number of public and private parties with concerns about this issue. The Governor's office submitted all comments received to DOE by March 1993, and is awaiting a response by DOE to its request for a meeting and consideration of certain modifications to the EA before it is completed. The following are some of the major public concerns about expanded pit storage:

- *The current lack of interim storage capacity throughout the Nuclear Weapons Complex may keep plutonium pits at Pantex.* Many citizens note that DOE itself has claimed it will be difficult to get approval to move plutonium pits from Pantex to another site because many facilities are old, deteriorated, or technically obsolete. Most DOE facilities were built between the 1940s and the mid-1960s with a strong emphasis on production activities and little emphasis on maintenance.

Upgrading to meet environmental, safety, and health standards is difficult and expensive. Shutdowns and prolonged outages are common at some sites. Often, DOE's "piecemeal improvement" approach has proved inadequate (27).

Furthermore, the criteria used for designing and building many of these facilities cannot meet today's standards for seismic design, fire protection, environmental safety, worker protection and safeguards, and security (27).

Other factors have also reduced the number of possible storage options. For instance, new approaches to risk and safety analyses have forced more stringent operating requirements. Therefore, the process of achieving compliance at old facilities is more difficult (27).

- *The lack of a clear timetable for siting and building a long-term plutonium storage facility has led to public skepticism about the estimated duration of pit storage at Pantex.* Many citizens, as well as State and local officials, believe that the lack of storage options for plutonium pits throughout the Nuclear Weapons Complex will result in Pantex becoming a de facto long-term storage facility (19,31).

The major reason for this concern is the uncertainty associated with the timetable for interim storage at Pantex. No definitive schedule has been developed to date regarding implementation of the Programmatic Environmental Impact Statement (PEIS) for reconfiguration of the Nuclear Weapons Complex. DOE has indicated in its draft Environmental Assessment that interim storage at Pantex will range from 6 to 10 years; however, some DOE officials have suggested that the storage period may last from 12 to 20 years or until a permanent storage site becomes available (23). To date, there is little coordination and integration between efforts to address interim storage of plutonium pits (EA) and long-term plutonium storage (PEIS).

- *Only limited information has been provided to the public about the analyses performed by DOE to support the conclusions reported in the Environmental Assessment.* Some have suggested that DOE release more data to the public. See box A-1 for comments obtained from the recent public review of DOE's Environmental Assessment.

OTHER FACTORS

Criticality Safety

Because plutonium is a fissile material, especially when present as plutonium-239, considerable attention must be paid during its handling to the prevention of criticality accidents.[3] According to a 1992 Technical Safety Appraisal Team report, the control of criticality accidents at Pantex is helped by the fact that the plutonium handled there is present in large metal components and weapons assemblies. To ensure effective criticality control, weapons assembly and dismantlement activities are required to comply with strict limits based on measurable control parameters. As part of the efforts to reduce the risk of criticality accidents, personnel are required to maintain fissionable materials inside containers when not in use. All storage and shipping containers used are certified as criticality safe. In addition, personnel must provide sufficient space (3 feet from center to center) between the nuclear weapons components and limit the number of components (not more than three in most cases) that can be left outside their containers in weapons assembly/disassembly areas (16,28).

Although most problems identified during the 1989 Tiger Team assessment of Pantex were found to have been addressed, a 1992 Technical Safety Appraisal Team review still indicated that the level of formal authority given to the ALARA (as low as reasonably achievable) coordinator was minimal regarding concurrence over programs that could either affect radiation levels or ensure that ALARA calculations and

[3] A criticality accident can occur due to the expansion of fissionable inventories or the removal of nonfissionable materials useful to prevent or terminate a criticality event. Such events can result in potentially lethal doses of neutron and gamma radiation to facility personnel, generate heat and fission products, and in certain situations, release radioactive materials to the outside environment. Unlike a nuclear reactor facility in which a criticality accident could release large quantities of fission products, some have estimated that the environmental releases caused by accidental criticality at DOE plutonium storage facilities would be negligible. Of much greater concern, a recent study concludes, would be the significant impact of a criticality accident on workers' health (22).

Box A-1—Expressed Public Concerns/Opinions About Plutonium Components at Pantex

- Limited information was provided about other locations as possible storage options (9,12).
- DOE estimates of the time workers would need to thoroughly inspect and inventory pits in bunkers could be longer than projected and result in higher radiation exposures of workers (10).
- The analysis of possible environmental impacts disregards current radioactive contamination of nearby soil (10).[1]
- Information on tornado analysis appears to exclude the high-frequency, potentially damaging tornadoes experienced in the Pantex area (20).
- The methodologies used for aircraft accident analyses have not been defined, and the data provided do not seem to reflect the density of air traffic (commercial and military) in the area or the proximity of Pantex to the airport (10,18)
- The basis for considering the storage of pits at Pantex for 6 to 10 years was not evaluated (9).
- The use of national security considerations to limit the information available for an open public debate is troubling.
- The decision to limit the Environmental Assessment to plutonium storage, without regard to other dismantlement activities at the site (2), is also a concern.
- DOE's efforts to integrate the Environmental Assessment with its PEIS for the Weapons Complex reconfiguration are lacking (9).
- The true degree of safety of nearby communities that can be guaranteed by DOE and State governments is unclear (6).

[1] Independent reviews of some of the limited public data available indicate that incorrect estimates of large aircraft crash rates and less than comprehensive estimates of air traffic in the region may have been used in the EA (11).

SOURCE: Office of Technology Assessment

procedures were followed. The review also stated that a certified external dosimetry system was lacking (28)

Plutonium Containers and Their Availability

The drums in which pits are contained for staging at bunkers, also known as AL-R8 containers, are made of carbon or stainless steel with an internal cage for holding or clamping the pits. To increase pit protection—and criticality safety—the drum is lined with 3 inches of cushioning and insulating material (24,28).

AL-R8 drums are characterized by excellent integrity and stability for a period of about 50 years. Identification and tracking are also efficiently provided. Knowledge of the performance of such containers over longer time periods (e.g., 100 years) is limited. Experts suggest that drum stability for longer periods be evaluated since the dynamic nature of plutonium in pit form may adversely impact the integrity and stability of the storage drums (5).

After having used AL-R8 containers to ship plutonium pits for nearly 30 years, DOE recently issued a policy abolishing this use and recommending instead the adoption of drums that more clearly comply with design criteria promulgated by the Department of Transportation and the Nuclear Regulatory Commission (17). One such design requirement is that there be an inner container vacuum system. DOE's newly designed pit shipping container satisfies this and all other criteria, but its cost (about $5,000) is significantly higher than that of AL-R8 drums (about $300) (16,26). Efforts to design a cheaper container are being carried out at the Savannah River Site and at Sandia National Laboratory (14). The limited availability of shipping containers could become a serious concern if a large number of pits require off-site transport for processing.

Accident Analysis

The EA accident analysis evaluated the potential of certain events to initiate accidents that could adversely affect the bunkers, or even worse, the drums in which the pits are contained.

DEFINITION OF INCREDIBLE EVENT

In the accident analysis, several events were dismissed as incredible and not considered in DOE's analyses of plutonium pit storage facilities at the Pantex Plant.[4] Examples of these include criticality events, internal/external explosions, internal fires, and meteor strikes. Other potential events, though evaluated, were also considered highly unlikely, including structural damage and/or penetration of the bunker by: 1) an explosion-generated missile from a hypothetical explosion of 50,000 pounds of high explosives stored at nearby bunkers; 2) a crash by a light, general aviation aircraft, with a weight of 3,500 pounds and an impact speed of 80 miles per hour; and 3) a tornado with speeds exceeding 140 pounds per square foot or about 340 miles per hour (24).

Events found to be most relevant to the proposed expanded pit storage at Pantex included:

- **Explosion hazards.** Explosions from other buildings at Pantex were evaluated to determine their potential to affect the plutonium pits stored inside bunkers. The results indicated that these pits would not be affected.
- **Structural hazards.** Analyses in the EA indicate that earthquakes, tornadoes, and external explosions would have no significant effect on the bunkers and their contents.
- **Operational accidents.** The operational event of most concern involved accidental puncture of a pit-containing drum, followed by crushing of the plutonium pit by a forklift moving at 5 miles per hour. Based on its analysis, DOE anticipates that the plutonium-containing dust escaping from the damaged container would: 1) have no immediate or long-term health effect on the worker involved in the accident; 2) cause only marginal radiation exposures to workers present in the immediate vicinity of the bunker where the accident took place; and 3) release no radiation to the public or the environment.
- **Aircraft crash.** The EA also contained an aircraft hazard analysis, which concluded that the probability of bunkers being impacted by an airplane was less than one in a million (24). According to some concerned citizens and public officials, DOE's decision to evaluate the risk of general aviation by using probability methodology designed to include all aircraft types may limit the validity of the results obtained (18) .
- **Accidents impacting groundwater sources (Ogallala Aquifer).** According to DOE estimates, no operating or accidental activities at Pantex were found capable of releasing plutonium at levels high enough to impact the underlying Ogallala Aquifer. A similar conclusion was reached in an independent study conducted by Los Alamos National Laboratory.
- **Internal/external fire hazards.** On the bases of the fire hazards analyses performed on Zone 4 (the highly secured storage area where the bunkers are located) as part of the Safety Analysis Report, DOE concluded that fires would have no impact on the bunkers or on the plutonium pit drums stored inside. The absence of combustible materials inside bunkers was also considered a major factor in further reducing fire risks (24).

CONCERNS ABOUT DOE'S ACCIDENT ANALYSIS

Reviewers have made the following comments about DOE's accident analysis:

> The term "incredible" connotes an event that is too improbable to be believed. Even though the DOE analysis states that an incredible event is one of very low probability, it also eliminates events that are categorized as such from further consideration. It would be more helpful and accurate if these events—especially when they have significant consequences—were included in a full analysis as "very low probability" events.

[4] DOE terms an event as "incredible" when its occurrence is too improbable or inconceivable due to the location of the facility, the environmental characteristics of the area surrounding the facility, and the nature of the materials used and operations conducted at the particular facility. Statistically, an event is called incredible when its probability of occurrence is calculated to be less than one chance in a million.

The Federal Aviation Administration's comment (21) on aircraft accidents makes the following point:

> The Aircraft Hazard Analysis Data on pages 6-5 through 6-8 and Appendix E of the Environmental Assessment prepared by the United States Department of Energy has no resemblance to the data provided by this office. Therefore, I am unable to comment on any information contained in the Assessment. For your information, the total aircraft operations for the Amarillo area in the CY 1992 was 91,800. Any further restrictions to flight or changes of airspace to the Pantex Prohibitive area would have an immediate and adverse impact on the utilization of Amarillo International Airport.

The Pantex Safety Analysis Report, on which the EA was based, was prepared by a DOE contractor under the guidance of the Albuquerque Field Office. It was reviewed by a technical safety review panel, all of whose members were from either DOE Defense Programs (DP) or the Albuquerque Field Office and approved by the Assistant Secretary for DP. The report was *not* prepared under the new safety analysis guidelines. No review by any internal DOE or external oversight group was done or sought. All accident analyses were accomplished with this Safety Analysis Report (4).

APPENDIX A REFERENCES

1. Alley, C., Mason & Hanger-Silas Mason Co., Inc., "Pantex ALARA Policy and Goals for 1992," a draft policy document prepared by Clyde Alley for distribution to Mason & Hanger's programs, undated.
2. Belisle, Mavis, The Peace Farm, personal communication to Roger Mulder, Director of Special Projects, Office of the Governor, Mar. 12, 1993.
3. Caskey, Bill, Sandia National Laboratories, New Mexico, briefing of Office of Technology Assessment staff, May 20, 1992.
4. Claytor, Richard A., Assistant Secretary for Defense Programs, U.S. Department of Energy, memorandum on "Interim Storage and Staging Activities at Pantex Zone 4 West" sent to Ted Dobry, Defense Programs (DP 6.2), Dec. 21, 1992.
5. Cunningham, Paul, Los Alamos National Laboratory, information provided at a briefing of the National Academy of Sciences Panel on Reactor-Related Options for Weapons Plutonium, Jan. 11, 1993.
6. Dees, Portia, personal communication to Roger Mulder, Director of Special Projects, Office of the Governor, State of Texas, Mar. 5, 1993.
7. "DOE Studying Plutonium Storage at Los Alamos, Savannah River," *Weapons Complex Monitor*, vol. 3, No. 28, June 22, 1992, pp. 18-19.
8. Fernandez, Tony, Sandia National Laboratories, New Mexico, information provided at a briefing of OTA staff, May 20, 1992.
9. Gattis, Beverly, Military Production Network, personal communication to Roger Mulder, Director of Special Projects, Office of the Governor, State of Texas, Mar. 12, 1993.
10. Hazlett, Margie K., personal communication to Roger Mulder, Director of Special Projects, Office of the Governor, State of Texas, Mar. 15, 1993.
11. Kimura, Chris, Fission Energy and Systems Safety Program, Lawrence Livermore National Laboratory, personal communication, Apr. 9, 1993.
12. Kleuskens, Tonya, Texas Nuclear Waste Task Force, letter to Roger Mulder, Director of Special Projects, Office of the Governor, Mar. 15, 1993.
13. Longenbaugh, Cindy, Weapons Programs Division, U.S. Department of Energy Albuquerque Operations Office, letter to Mark Brown, Office of Technology Assessment, Nov. 4, 1992.
14. Longmire, Paul, Sandia National Laboratories, briefing before Office of Technology Assessment staff, Nov. 11, 1992.
15. McBride, J., "DOE Assessing Pantex Dismantlement," *Amarillo Daily News*, June 6, 1992.
16. McLaughlin, T., Group Leader, Nuclear Criticality Safety Program, Los Alamos National Laboratory, personal communication, Mar. 2, 1993.
17. McLaughlin, T., Group Leader, Nuclear Criticality Safety Program, Los Alamos National Laboratory, information provided during a briefing of OTA staff, Mar. 10, 1993.
18. Miller, Alison A., Air Quality Assessment Program, Texas Air Control Board, State of Texas, letter to Roger Mulder, Director of Special Projects, Office of the Governor, Mar. 1, 1993.
19. Olinger, Nancy, Office of Attorney General, State of Texas, personal communication to Kenneth Pusateri, Defense Nuclear Facilities Safety Board, Sept. 18, 1992.

20. Osborne, Judy, personal communication to Roger Mulder, Director of Special Projects, Office of the Governor, State of Texas, Mar. 2, 1993.
21. Schulke, C.R., Air Traffic Manager, Federal Aviation Administration, U.S. Department of Transportation, letter to Joseph A. Martillotti, Bureau of Radiation Control, Pantex Special Project Coordinator, Jan. 28, 1993.
22. SCIENTECH, *An Assessment of Criticality Safety at the Department of Energy Rocky Flats Plant, Golden, Colorado; July-September 1989*, SCIE-DOE-201-89, Contract No. DE-AC01- 88EH79081/ Task 20, September 1989.
23. "Texas Questioning Plutonium Storage Configurations at Pantex," *Weapons Complex Monitor*, Dec. 23, 1992.
24. U.S. Department of Energy, Albuquerque Operations, Amarillo Area Office, "Environmental Assessment for Interim Storage of Plutonium Components at Pantex," predecisional document, DOE/EA- 0812, December 1992.
25. U.S. Department of Energy, memorandum Summary and Action Items, DOE Executive Management Team for Dismantlement Meeting 92-6, Aug. 5-6, 1992.
26. U.S. Department of Energy, memorandum Summary and Action Items, DOE Executive Management Team for Dismantlement Meeting 92-5, June 29, 1992.
27. U.S. Department of Energy, *Nuclear Weapons Complex Reconfiguration Study*, DOE/DP-0083 (Washington, DC: U.S. Department of Energy, January 1991).
28. U.S. Department of Energy, Office of Defense Programs, Office of Inspections, "Technical Safety Appraisal of the Pantex Plant, Amarillo, Texas," March 1992.
29. U.S. Department of Energy, Savannah River Field Office, Environmental Division, "Environmental Assessment: Storage of plutonium Metal in Building 247-F Vault," DOE/EA-0497, February 1992.
30. West, George T., Mason & Hanger-Silas Mason Co., Inc., Pantex Plant, Amarillo, Texas, "United States Nuclear Warhead Assembly Facilities (1945-1990)," March 1991.
31. Zamora, Thomas, Friends of the Earth, Washington, DC, personal communication, Nov. 9, 1992.

Appendix B: United States and Russia Joint Understanding on Strategic Offensive Arms

The President of the United States of America and the President of the Russian Federation have agreed to substantial further reductions in strategic offensive arms. Specifically, the two sides have agreed upon and will promptly conclude a Treaty with the following provisions:

1. Within the 7-year period following entry into force of the START Treaty, they will reduce their strategic forces to no more than:
   a. an overall total number of warheads for each between 3,800 and 4,350 (as each nation shall determine) or such lower number as each nation shall decide.
   b. 1,200 MIRVed ICBM warheads.
   c. 650 heavy ICBM warheads.
   d. 2,160 SLBM warheads.

2. By the year 2004 (or by the end of the year 2000 if the United States can contribute to the financing of the destruction or elimination of strategic offensive arms in Russia), they will:
   a. reduce the overall total to no more than a number of warheads for each between 3,000 and 3,500 (as each nation shall determine) or such lower number as each nation shall decide.
   b. eliminate all MIRVed ICBMs.
   c. reduce SLBM warheads to between no more than 1,700 to 1,750 (as each nation shall determine).

3. For the purpose of calculating the overall totals described above:
   a. The number of warheads counted for heavy bombers with nuclear roles will be the number of nuclear weapons they are actually equipped to carry.
   b. Under agreed procedures, heavy bombers not to exceed 100 that were never equipped for long-range nuclear ALCMs and that are reoriented to conventional roles will not count against the overall total established by this agreement.
      i. Such heavy bombers will be based separately from heavy bombers with nuclear roles.
      ii. No nuclear weapons will be located at bases for heavy bombers with conventional roles.
      iii. Such aircraft and crews will not train or exercise for nuclear missions.
      iv. Current inspection procedures already agreed in the START Treaty will help affirm that these bombers have conventional roles. No new verification procedures are required.

4. The reductions required by this agreement will be carried out by eliminating missile launchers and heavy bombers using START procedures, and, in accordance with the plans of the two sides, by reducing the number of warheads on existing ballistic missiles other than the SS-18. Except as otherwise agreed, ballistic missile warheads will be calculated according to START counting rules.

NOTE: This appendix is a reprint of the text of the June 1992 Joint Understanding Between the United States and Russia on Reductions in Strategic Offensive Arms Incorporated in the START II Treaty, signed January 1993.

5. The two Presidents directed that this agreement be promptly recorded in a brief Treaty document which they will sign and submit for ratification in their respective countries. Because this new agreement is separate from but builds upon the START Treaty, they continue to urge that the START Treaty be ratified and implemented as soon as possible.

Done at Washington, this 17th day of June 1992, in two copies, each in the English and Russian languages, both texts being equally authentic.

Appendix C: United States and Russian Agreement on the SSD Program

The United States of America and the Russian Federation, hereinafter referred to as the Parties, desiring to facilitate the safe and secure transportation and storage of nuclear, chemical, and other weapons in the Russian Federation in connection with their destruction, intending to build upon the framework for cooperation set forth in the Agreement Between the Government of the United States of America and the Government of the Russian Federation Regarding Cooperation to Facilitate the Provision of Assistance of April 4, 1992, have agreed as follows:

ARTICLE I

The Parties shall cooperate in order to assist the Russian Federation in achieving the following objectives:

a. the destruction of nuclear, chemical, and other weapons;
b. the safe and secure transportation and storage of such weapons in connection with their destruction; and
c. the establishment of additional verifiable measures against the proliferation of such weapons that pose a risk of proliferation.

ARTICLE II

1. The Parties, through their Executive Agents, shall enter into implementing agreements as appropriate to accomplish the objectives set forth in Article I of this Agreement. The implementing agreements shall include, inter alia:
   a. a description of the activities to be undertaken;
   b. provisions concerning the sequence of activities;
   c. provisions concerning access to material, training or services provided at sites of their use, if possible, for monitoring and inspection and
   d. other provisions as appropriate.
2. In Case of any inconsistency between the Agreement and any implementing agreements, the provisions of this Agreement shall prevail.

ARTICLE III

Each Party shall designate an Executive Agent to implement this Agreement. For the United States of America, the Executive Agent shall be the Department of Defense. For the Russian Federation, with respect to nuclear weapons, the Executive Agent shall be the Ministry of Atomic Energy.

ARTICLE IV

Except as otherwise provided in this Agreement or in an implementing agreement, the terms of this Agreement shall apply to all material, training or services provided in accordance with this Agreement or implementing agreements, and to all related activities and personnel.

NOTE: This appendix is a reprint of the text of the "Umbrella" Agreement Between the United States and Russia Concerning Assistance for Weapons Dismantlement in Russia (The "Safe, Secure, Dismantlement" (SSD) Program), signed June 1992.

ARTICLE V

1. The Russian Federation shall facilitate the entry and exit of employees of the Government of the United States of America and contractor personnel of the United States of America into and out of the territory of the Russian Federation for the purpose of carrying out activities in accordance with this Agreement.
2. Aircraft and vessels, other than regularly scheduled commercial aircraft and vessels, used by the United States of America in connection with activities pursuant to this Agreement in the Russian Federation shall, in accordance with international law, be free of customs inspections, customs charges, landing fees, navigation charges, port charges, tolls, and any other charges by the Russian Federation, or any of its instrumentalities.
3. If an aircraft other than a regularly scheduled commercial aircraft is used by the United States of America for transportation to the Russian Federation, its flight plan shall be filed in accordance with the procedures of the International Civil Aviation Organization applicable to civil aircraft, including in the remarks section of the fight plan confirmation that the appropriate clearance has been obtained. The Russian Federation shall provide parking, security protection, servicing, and fuel for aircraft of the United States of America.

ARTICLE VI

Unless the written consent of the United States of America has first been obtained, the Russian Federation shall not transfer title to, or possession of, any material, training or services provided pursuant to this Agreement to any entity, other than an officer, employee or agent of a Party to this Agreement and shall not permit the use of such material, training or services for purposes other than those for which it has been furnished.

ARTICLE VII

1. The Russian Federation shall, in respect of legal proceedings and claims, other than contractual claims, hold harmless and bring no legal proceedings against the United States of America and personnel, contractors, and contractors' personnel of the United States of America, for damage to property owned by the Russian Federation, or death or injury to any personnel of the Russia Federation, arising out of activities pursuant to this Agreement.
2. Claims by third parties, arising out of the acts or omissions of any employees of the United States of America or contractors or contractors' personnel of the United States of America done in the performance of official duty, shall be the responsibility of the Russian Federation.
3. The provisions of this Article shall not prevent the Parties from providing compensation in accordance with their national laws.
4. The Parties may consult, as appropriate, on claims and proceedings under this Article.
5. Nothing in this Article shall be construed to prevent legal proceedings or claims against nationals of the Russian Federation or permanent residents of the Russian Federation.

ARTICLE VIII

The activities of the United States of America under this Agreement are subject to availability of appropriated funds.

ARTICLE IX

Employees of the Government of the United States of America present in the territory of the Russian Federation for activities related to this Agreement shall be accorded privileges and immunities equivalent to that accorded administrative and technical staff personnel in accordance with the Vienna Convention on Diplomatic Relations of April 18, 1962.

ARTICLE X

1. The United States of America, its personnel, contractors, and contractors' personnel shall not be liable to pay any tax or similar charge by the Russian Federation or any of its instrumentalities on activities undertaken in accordance with this Agreement.
2. The United States of America, its personnel, contractors, and contractors' personnel may import into, and export out of, the Russian Federation any equipment, supplies, material or services required to implement this Agreement. Such importation and exportation of articles or services shall not be subject to any license, other restrictions, customs, duties, taxes or any other charges or inspections by

the Russian Federation or any of its instrumentalities.

ARTICLE XI

In the event that a Party awards contracts for the acquisition of articles and services, including construction, to implement this Agreement, such contracts shall be awarded in accordance with the laws and regulations of that Party. Acquisition of articles and services in the Russian Federation by or on behalf of the United States of America in implementing this Agreement shall not be subject to any taxes, customs, duties or similar charges by the Russian Federation or its instrumentalities.

ARTICLE XII

The Russian Federation shall take all reasonable measures within its power to ensure the security of material, training or services provided pursuant to this agreement and shall protect them against seizure or conversion.

ARTICLE XIII

Upon request, representatives of the Government of the United States of America shall have the right to examine the use of any material, training or other services provided in accordance with this Agreement, if possible at sites of their location or use, and shall have the right to inspect any and all related records or documentation during the period of this Agreement and for 3 years thereafter. These inspections shall be carried out in accordance with procedures to be agreed upon by the Parties.

ARTICLE XIV

This Agreement shall enter into force upon signature and shall remain in force for 7 years. This Agreement may be amended or extended by the written agreement of the Parties and may be terminated by either Party upon ninety days written notification to the other Party of its intention to do so. Notwithstanding the termination of this Agreement or the implementing agreements, the obligations of the Russian Federation in accordance with Articles VI, VII, IX, X, XII of this Agreement shall continue to apply without respect to time, unless otherwise agreed in writing by the Parties.

Done at Washington this 17th day of June 1992, in two copies, each in the English and Russian languages, both texts being equally authentic.

Appendix D: United States and Russian Agreement on the Disposition of Uranium from Nuclear Weapons

The Government of the United States of America and the Government of the Russian Federation, hereinafter referred to as the Parties, desiring to arrange the safe and prompt disposition for peaceful purposes of highly enriched uranium extracted from nuclear weapons resulting from the reduction of nuclear weapons in accordance with existing agreements in the area of arms control and disarmament.

Reaffirming their commitment to ensure that the development and use of nuclear energy for peaceful purposes are carried out under arrangements that will further the objectives of the Treaty on the Non-Proliferation of Nuclear Weapons, affirming their commitment to ensure that the nuclear material transferred for peaceful purposes pursuant to this Agreement will comply with all applicable non-proliferation, physical protection, nuclear material accounting and control, and environmental requirements, have agreed as follows:

ARTICLE I

Purpose

The Parties shall cooperate in order to achieve the following objectives:

1. The conversion as soon as practicable of highly enriched uranium (HEU) extracted from nuclear weapons resulting from the reduction of nuclear weapons pursuant to arms control agreements and other commitments of the Parties which is currently estimated at approximately 500 metric tons in the Russian Federation, having an average assay of 90 percent or greater of the uranium isotope 235 into low enriched uranium (LEU) for use as fuel in commercial nuclear reactors. For purposes of this Agreement, LEU shall mean uranium enriched to less than 20 percent in the isotope 235; and
2. The technology developed in the Russian Federation for conversion of HEU resulting from the reduction of nuclear weapons in the Russian Federation may be used for conversion of United States HEU in the United States of America; and
3. The establishment of appropriate measures to fulfill the non-proliferation, physical protection, nuclear material accounting and control, and environmental requirements of the Parties with respect to HEU and LEU subject to this Agreement.

ARTICLE II

Purpose

1. The Parties, through their Executive Agents, shall within 6 months from entry into force of this Agreement seek to enter into an initial implementing contract to accomplish the objectives set forth in Article I of this Agreement. The Parties may

NOTE: This appendix is a reprint of the text of the Agreement Between the United States and Russia Concerning the Purchase of Highly Enriched Uranium Extracted from Russian Warheads, signed February 1993.

conclude additional implementing contracts or agreements pursuant to this Agreement, as required. For any purchase, the Executive Agents shall negotiate terms (including price), which shall be subject to approval by the Parties.

2. It is the intent of the Parties that the initial implementing contract shall provide for, inter alia:
   - i. The purchase by the United States Executive Agent of LEU converted from HEU at facilities in the Russian Federation and sale of such LEU for commercial purposes. The United States will provide information to the Russia Federation on all commercial disposition of such LEU;
   - ii. Initial delivery of LEU converted from HEU extracted from nuclear weapons resulting from the reduction of nuclear weapons pursuant to arms control agreements and other commitments of the Parties by October 1993, if possible;
   - iii. Conversion of no less than 10 metric tons having an average assay of 90 percent or greater of the uranium isotope 235 in each of the first five years, and, in each year thereafter, conversion of no less than 30 metric tons of HEU having an average assay of 90 percent or greater of the uranium isotope 235; however, specific amounts will be stipulated in the first and subsequent implementing contracts or agreements;
   - iv. The participation of the United States private sector and of Russian enterprises;
   - v. The allocation among the United States of America, private sector firms of the United States of America, the Russian Federation, and Russian enterprises of any proceeds or costs arising out of activities undertaken pursuant to any implementing contracts;
   - vi. The use by the Russian side of a portion of the proceeds from the sale of LEU converted from HEU for the conversion of defense enterprises, enhancing the safety of nuclear power plants, environmental clean-up of polluted areas and the construction and operation of facilities in the Russian Federation for the conversion of HEU to LEU;
   - vii. By agreement of the Parties an equivalent amount of HEU can substitute for the corresponding amount of LEU planned for purchase by the United States Executive Agent.

ARTICLE III

Executive Agents

Each party shall designate an Executive Agent to implement this Agreement. For the United States side, the Executive Agent shall be the Department of Energy. For the Russian side, the Executive Agent shall be the Ministry of the Russia Federation of Atomic Energy. After consultation with the other Party, either Party has the right to change its Executive Agent upon 30-days written notice to the other Party. If a governmental corporation is established under United States law to manage the uranium enrichment enterprise of the Department of Energy, it is the intention of the United States Government to designate that corporation as the Executive Agent for the United States side.

Priority of Agreement

In case of any inconsistency between this Agreement and any implementing contracts or agreements, the provisions of this Agreement shall prevail.

Additional Measures

1. The Executive Agent of the Russian Federation shall ensure that the quality of LEU derived from HEU subject to this agreement is such that it is convertible to LEU usable in commercial reactors. Specifications shall be agreed upon in the process of negotiating the initial and subsequent implementing contracts.
2. The conversion of HEU subject to this Agreement shall commence as soon as possible after the entry into force of the initial implementing contract.
3. The Parties shall, to the extent practicable, seek to arrange for more rapid conversion of HEU to LEU than that provided for in Article II (2) (iii).
4. The United States of America shall use LEU acquired pursuant to this Agreement and its implementing contracts and agreements, when subject to United States jurisdiction and control, for peaceful purposes only.

5. LEU acquired by the United States of America pursuant to this Agreement, and implementing contracts and agreements related to it, shall be subject to safeguards in accordance with the November 18, 1977, Agreement Between the United States of America and the International Atomic Energy Agency (IAEA) for the Application of Safeguards in connection with the Treaty on the Non-Proliferation of Nuclear Weapons.
6. The Parties shall maintain physical protection of HEU and LEU subject to this Agreement. Such protection shall, at a minimum, provide protection comparable to the recommendation set forth in IAEA document INFCIRC/225/REV.2 concerning the physical protection of nuclear material.
7. If the Parties enter into an agreement for cooperation concerning the peaceful uses of nuclear energy, nuclear material acquired by the United States of America pursuant to this Agreement and its implementing contracts and agreements, when subject to United States jurisdiction or control, shall be subject to the terms and conditions of that Agreement for cooperation.
8. The activities of the United States Government under this Agreement, or any implementing contract or agreement, shall be subject to the availability of United States Government funds.
9. In the event the United States Government does not have funds available for implementation of this Agreement, the Executive Agent of the Russian Federation reserves the option to obtain funding for implementation of this Agreement from any private United States company.
10. Prior to the conclusion of any implementing contract, the Parties shall establish transparency measures to ensure that the objectives of this Agreement are met, including provisions for nuclear material accounting and control and access, from the time that HEU is made available for conversion until it is converted into LEU. Specific transparency measures shall be established in the same time frame as the negotiation of the initial implementing contract, and shall be executed by a separate agreement.
11. Prior to the conclusion of any implementing contract, the Parties shall agree on appropriate governing provisions for entry and exit, liability, and status of personnel, exemptions for taxes and other duties, and applicable law.
12. The Executive Agent of the United States of America shall use the LEU converted from HEU in such a manner so as to minimize disruptions on the market and maximize the overall economic benefit for both Parties. This Agreement shall have no effect on contracts between Russian enterprises and United States companies for the delivery of uranium products which are currently in force and consistent with United States and Russian law.
13. This Agreement places no limitations on the right of the Russian Federation to dispose of LEU derived from HEU extracted from nuclear weapons resulting from the reduction of nuclear weapons pursuant to arms control agreements and other commitments of the Parties beyond the specific commitments set forth herein.

Entry into Force, Duration and Amendments

1. This Agreement shall enter into force upon signature and shall remain in force until the full amount of HEU provided for in paragraph 1 or Article I is converted into LEU, delivered, and supplied to commercial customers.
2. Each Party may propose amendments to this Agreement. Agreed amendments shall enter into force upon signature and shall remain in force so long as this Agreement remains in force.
3. Each Party shall have the right to terminate this Agreement upon 12 months written notification to the other Party.

Done at Washington this 18th day of February 1993, in duplicate in the English and Russian languages, both texts being equally authentic.

Appendix E: Labor and the Mission Change Within DOE

This appendix presents comments by labor leaders on worker experiences with health and safety priorities at selected Department of Energy (DOE) weapons facilities.[1] Facilities included those that have recently been affected by a shift in mission from weapons production to weapons dismantlement and the storage and disposition of bomb parts. The comments are from telephone and personal interviews by an anthropologist with organized labor leaders at the Oak Ridge Y-12 Plant, the Pantex Plant, and the Savannah River Site. Comments are also included from the Oak Ridge K-25 Plant, a facility whose mission has been formally changed from weapons production to environmental restoration. Although the interview pool was small and uneven in terms of craft representation, the individuals surveyed are key figures in the work force. Interviewees included those holding leadership positions, as well as workers with responsibility for health and safety issues.

Questions were asked about workers' perceptions of DOE's efforts to change its culture from a strict production orientation to one that emphasizes the protection of health, safety, and the environment. Topics addressed included DOE oversight; policies and procedures for protecting the environment, safety, and health; training; resolution of complaints; health surveillance; and relationships among labor, contractors, and DOE. This appendix, however, departs from a cataloging of interview comments and, instead, describes labor issues that may impinge on the success of further transformations within DOE.

HEALTH AND SAFETY IN THE CONTEXT OF LABOR-MANAGEMENT RELATIONS

■ Changing Concepts of Work

Under the direction of former Secretary, James D. Watkins, DOE instituted an environmental, safety, and health agenda that has begun to translate into a different experience of work for the labor force (7). Worker comments on the revised regulations, procedures, and standards accompanying DOE's new agenda are suggestive of a growing disjuncture between old rhythms of work and new ones. Although the following worker comments indicate that change has begun, it remains to be seen whether DOE's commitment of social and economic resources to this agenda is adequate.

"REAL WORK" VERSUS HEALTH AND SAFETY

Remarks made by some contractor employees on the change in "culture" at DOE are reminders of the notion of work developed under an environment of production at all costs. Production schedules determined what was valued and experienced as real work and what essentially kept people in a job.

Production was "born and bred" into the work force over the years, as one worker describes it, and recent changes go against this production "mindset" that

[1] The interviews were conducted, and summaries of comments prepared, by Monica Schoh-Spana, a contractor to the Office of Technology Assessment. The interviews were conducted during the winter of 1992-93.

developed over time. The new focus on adherence to standards in production, increased attention to waste streams, an upgraded security system, and scrutiny of operations by oversight groups are at odds with production employees' prior work histories. How extraordinary these changes appear varies among different generations of workers, according to workers from this facility. The average age of the work force at one facility is more than 40 years. Younger workers are learning more from the start about the environment and the appropriate way of disposing of things. In contrast, older workers recall a past when they had permission to "throw things outside" with little regard for the environment.

The new culture that emphasizes adherence to standards clashes with the routines and knowledge of technicians cultivated and valued under the old culture of production. In the past, explains a worker, the message was "get the product out, regardless of standards." Supervisors and managers condoned going against standards if it meant that the product was moving. This is no longer the case, proposes another contact. DOE, managers, and workers are clear on the importance of adherence to standards on the line. If a technician has a problem with a standard, he will not circumvent or ignore it. He will shut down the job and confer with an engineer or other appropriate personnel about the problem. The transition on the line, however, has been a difficult one for some workers. In following standards strictly, workers cannot use the skills and shortcuts they developed to get a job done when supervisors and managers pushed production at all costs. Explains one worker, people have been working here for more than 20 years, and now they are asked to change the way they do their jobs overnight.

The drastic change in experience, however, has been difficult not only for floor-level employees but also for line managers. At one facility, notes one interviewee, line supervisors have retired because they were not able to cope with the significant transformations in operations. Older workers and supervisors at this facility have often questioned the recent modifications with statements such as "I've been doing this job for 40 years, why do I need to change?"

Workers' remarks on the personal experience of this culture change bring into relief DOE's historical emphasis on production. How work was defined, valued, and experienced before DOE's reorganization still, in large part, sets the terms for work today. One contact explains that a shift away from production has given people an opportunity "to catch up" with environmental, safety, and health concerns. He jokes that with less work to do, everyone at the facility can put more of an emphasis on such matters. Work, his remark suggests, is really about production, and safety, health, or environmental concerns are in effect an overlay to work. These matters are considered a sort of luxury activity—something standing outside of real work. When the orientation toward the environment, safety, and health was first initiated, new regulations and standards came pouring in. The newness, inconsistency, and frequent changes in standards and procedures have made work difficult, explain two contacts at a facility. Rules and regulations "slow down the work process," according to one contact, and therefore workers think it is a waste of time to try to meet standards. Another contact also recounts that some new procedures tend "to slow down the work," which makes some people feel "less productive." This individual has conveyed to his coworkers the importance of doing the job correctly, according to standards. If the procedures lengthen the time frame of a job, then they just need to accept it, he suggests. Once a job has been "proceduralized," he says, it isn't the same job any more.

Experiences with new procedures related to radiation control and security convey more concretely the clash of the two different work processes. A worker from one facility relates that it is taking longer to get in and out of a work area because of the new requirements in personal monitoring for contamination. The monitoring turns into a hassle if someone goes in and out of an area several times a day. Workers become frustrated, he proposes, because they feel that they are "spending more time exiting an area than being on the job." A worker at another facility recounts the aggravation that has accompanied the installation of a more rigorous security system. He argues that the new equipment is poor, and it may take some people up to five tries before they are cleared to enter a work area. This can be extremely irritating, and people may lose their tempers. As one contact proposes, "All in all, people want to get their work done."

The physical layout of work space at some facilities and the newly mandated radiation control practices may translate into a real inconvenience and source of stress for workers, as may a new security system. That aspect of some of the recent changes cannot be ignored. It is helpful, nonetheless, to listen to the way in which workers talk about their experiences with the new procedures. They provide a commentary about the new rhythm of work demanded by these procedures. Furthermore, they reflect a frustration about how such procedures can get in the way of getting a job done, a job that was previously measured strictly in terms of productivity. If the procedures and the standards are indeed a means to acquire a safer workplace and a healthier environment overall, then hopefully the work force will in time develop new routines and habits. A work culture impelled by considerations of the environment, safety, and health will become second nature, much in the same way that a production-driven one did.

WORKER SAFETY VERSUS WORKER HEALTH

As scrutiny of the Nuclear Weapons Complex has increased and evidence of earlier disregard for the environment, safety, and health has come to light, definitions of how safe is safe and who gets to decide are still evolving. Historically, DOE and its contractors operated outside the realm of regulatory agencies such as the Occupational Safety and Health Administration (OSHA) and the Environmental Protection Agency (EPA), which set limitations on private industry. Workers have, however, begun to feel the effects of new health and safety standards. One of former Secretary Watkins's 1989 initiatives, for instance, was to redirect DOE from adoption of OSHA standards (a stated policy) to compliance with OSHA standards (an implemented policy)(7). This new emphasis has had an unsettling effect on the work force. Individuals from "outside" the facility, persons without firsthand knowledge of facility practices, are questioning the plant's safety. Having their work and safety records investigated by new agents is an uncomfortable experience for the workers. At the same time, workers are asking themselves what the new push is about. If DOE and its contractors are concerned now, why weren't they concerned before?

In addressing the new focus on "safety," workers at two facilities cite their facility's safety record, both figurative and literal, as defense of a long-standing commitment to safety, not just a recent one. The point here is not to dispute what may be admirable safety records, but to note that the records are used to gauge whether or not a priority on safety exists. In fact, a review of the Nuclear Weapons Complex by the National Research Council characterized its occupational safety record as "excellent" relative to the private sector's track record in lost workday incidence (2). On the other hand, OSHA, in its review of DOE's occupational safety and health program, identified instances in which contractors kept injury and illness rates artificially low (5). Some workers interviewed also propose that written accounts are not true representations of the incidence of injury and illness at DOE facilities.

Whether or not incident records are valid appraisals of health and safety at DOE facilities, the "good safety record" is invoked by the agency and even by some workers as the sign that safety has been a priority. The accident or catastrophe—the "life or limb" problem, as an interviewee describes it—may be the primary measure of disregard for the health and safety of workers, but perhaps at the price of obscuring more long-term threats to their health. Workers draw upon safety records in defense of their accomplishments and long-term concern for such matters, because this information is definitive and available. With regard to environmental, safety, and health protection within the Weapons Complex, one means of judging safety and health—with a short-term orientation—has been fostered at the expense of another, which reflects a long-term view. Review of DOE-sponsored or contracted epidemiological research concluded that the agency's pronouncement that the health of workers has been fully protected and that "there are no excess risks of disease and death in the nuclear weapons work force" could not be substantiated (3).

The ideal new culture will hopefully develop among workers an awareness of and precautions against both immediate and distant health threats. An important caveat here is that workers have not been unconscious of the threat that their work poses to their overall health. They have pushed for adequate health monitoring and appropriate analysis of worker epidemiological data. Their efforts have, however, been impeded by the stance of DOE.

FROM SECRECY TO OVERSIGHT

The workplace at one dismantlement facility has been transformed through the culture change from an insulated environment to one now subject to the intense scrutiny of oversight groups. This aspect of operations since the Watkins mandate has been one of the most burdensome for some workers. One production technician describes the waves of oversight in terms of an affront to the competence, dedication, and safety consciousness of the work force, which has a lengthy history at the facility. He underscores that members of the work force know how to do their jobs. They established this unique plant, and they consider themselves "the best of the best." Since the Watkins mandate, "outsiders" have come in and told the work force to do what it has always done at the facility—work safely. Another worker shares this concern with the increasing oversight of groups such as State agencies, and the implication that the work force has somehow disregarded environmental concerns.

Workers' experience with oversight goes beyond the feeling that outsiders are judging their work and questioning workers' abilities to achieve safe operations. Their frustration also includes the disruptive influence that oversight activities have on actual line operations. Work has been transformed from a relatively isolated experience with technicians working in the bay area on the buddy system into a spectacle with a total of 10 to 15 people in the bay at one time. This crowding, plus the distractions that the oversight people create, can affect their work, proposes a line technician. He worries about the potentially hazardous consequences of oversight, a process implemented to protect environmental, safety, and health concerns. Oversight is becoming a permanent feature of operations at many DOE facilities, and many different parties have a stake in defining what is meant by safe. While workers adjust to a different work environment—one open to scrutiny—they must also have confidence that this observation of their activities is indeed a measure that promotes safety.

The Environment, Safety, and Health—A New Culture?

The frustration that has accompanied the new requirements of the workplace within the Nuclear Weapons Complex is not, however, simply a matter of adjusting from one set of work expectations and habits to another. The consternation of the work force goes beyond adjustment to the different pace of work that radiation controls create or to the presence of other people who have a stake in defining "safe" operations. One should not explain away the skepticism or misgivings toward particular aspects of the culture change by relying on a psychological frame of reference (i.e., pointing out the discomfort that naturally attends a change in routine). Nor can one simply hope that time is the only ingredient missing for the culture change to take root.

WHEN AND FOR WHOM?

Some workers, for instance, question the sincerity of the recent emphasis on environmental, safety, and health matters because the push for production still exists, particularly at lower levels of authority. While top managers are committed to the new culture, lower-level managers (from mid-management down to line management) have production schedules that they must meet, explains a line technician. They have to answer to top managers about production, and some of the production schedules make it almost impossible to protect worker health and safety. While the culture change has made some headway, workers still receive mixed messages from management: health and safety, yes, but don't forget production. These messages also vary across departments, the line technician notes. Some foremen shut down if there is a problem, whereas others push people to produce regardless. There has been a shift, however, from past emphases. In the past, every level of the organization was oriented to production—from top management to mid-management to the worker. Now, some tradeoffs are being made between production and worker health and safety.

The continued push for production at the expense of health and safety in some quarters may invalidate the environmental, safety, and health agenda recently put forth by DOE and undermine the new sense of work developing on the floor. If workers feel that management is still emphasizing production, and if they understand this as the bottom line in keeping a job, new regulations and procedures that inhibit getting a job done may be experienced as an undue burden. This situation will continue in the face of irreconcilable DOE orders—some that stress production and others that stress health and safety matters. If workers are inclined to want to get on with their work, it may not be out of a fondness for or familiarity with some notion of past work. The Advisory Committee on Nuclear

Facility Safety has suggested that "production" is the default practice when workers are presented with conflicting and unclear directions (1).

WHOSE CULTURE CHANGE IS THIS?

While the perpetuation of production priorities at some managerial levels may inhibit assimilation of the new focus by the work force, a further hindrance to worker engagement in the new culture is the manner of its implementation. The punitive aspects of recent changes, as well as the limited inclusion of labor in discussions and decisions regarding this new focus, demonstrate the same hierarchical relations as in the Nuclear Weapons Complex that disregarded concerns for the environment, safety, and health. As one contact describes it, there are two ways you can make health and safety a priority. You can come in and dictate a new program or you can include input from the field. Employees, he argues, have valuable insight into making health and safety happen. People in the field possess skills and knowledge that can make health and safety into a policy that works on the ground.

The experience of environmental, safety, and health concerns as another managerial edict derives in part from the penalties that have accompanied recent changes in the workplace. A worker proposes that the "scare tactics" accompanying the new principle of strict adherence to standards have made the transition difficult. He recounts the following incident as an example. Members of an oversight group asked questions and created disturbances while technicians were working on a unit. During this high-pressure scrutiny, the workers missed some minor steps and were out of compliance with standards. In reaction to this violation, the plant manager threatened to fire all four workers, contrary to existing contract language.

A contact at another facility also relates how DOE "has beaten people over the head" with safety, which may affect the success of the new culture. In November 1992, a fatality occurred at a Weapons Complex site. In response, DOE implemented a blanket policy increasing the suspension period normally following an incident, from 1 to 2 days without pay. This measure was inappropriate, argues this worker, because of the current reprimand policy under which the contractor was already working as indicated by the safety record. From February to November of that same year, the rate of incidents had steadily decreased. Whereas no one disputes the value of appropriate penalties and rewards for inculcating the new value placed on the environment, safety, and health, the punitive nature of work reorganization may undermine workers' investment in the agency's new priorities.

The issue of penalty also raises questions of culpability and workers' concerns that liability for these matters rests squarely on their shoulders. A worker evaluating the recent imposition of rules and standards notes that technicians are taken aback by their sheer volume. People have to sign off on a procedure stating that they understand it, he explains. He worries if the fact that he signed off on procedures will come back to haunt him, that is, whether that documentation will serve as proof for his need of a reprimand should an incident occur. The assimilation of new practices and priorities should entail a sense of accountability. However, if the process for developing accountability entails pushing liability to the lowest levels of the work force, there is a problem: not only in labor's hesitancy to accept new environmental, safety, and health priorities for fear that it will become the "fall guy" for problems whose source it cannot control, but also in the possibility that contractors will use a system of individualized worker accountability (through documentation, for instance) as indemnity against their own accountability in an incident.

Workers constitute one of the purported beneficiaries of the new emphasis on safety and health; yet they have been given a relatively small role in implementing and evaluating the changes under way. DOE, which historically has had virtually no presence on site, has begun to exert a strong influence on day-to-day operations of the Weapons Complex facilities with its new agenda. Although the work force has been told to change the pace and content of jobs, there is no formal process by which workers can respond to the changes asked of them. The top-down fashion in which DOE has instituted these new priorities raises a question in the mind of some workers of whether the agency is more interested in demonstrating to its critics and detractors that it is doing *something* (i.e., "covering their tails" to quote one worker) than in fostering true protections for the environment, safety, and health. If it is committed to a real culture change, then DOE should develop a system for conferring directly with labor to learn through its experience. The Advisory Committee on Nuclear Facility Safety has already pointed out the importance of developing channels for effective communication from the floor level up, in order to strengthen a "safety culture" (1).

Frustrating labor's incorporation of the new work practices that comprise the culture change has been a lack of dialogue with DOE and management regarding the reconfiguration of work, in particular, and the redirection of the facilities and the Weapons Complex, in general. To resolve the tension between the new emphasis on adherence to standards and the skills of the work force, one worker suggests a process for evaluation of standards that involves the technician. Although some standards are appropriate, he proposes, others may not be. The process for evaluating new standards should include technicians, and not involve merely supervisors and engineers who hand technicians the final word on a standard. Another worker notes that the union will be participating in a future procedure validation team—an improvement in incorporating the work force into the culture change. A contact at a different facility also relates that the struggle for worker input is constant and cites the exclusion of labor in the development of DOE's hoisting and rigging manual as a recent example.

Although some facilities are still moving in the direction of improving floor-level contributions to work reorganization, workers at another facility have had the opportunity to participate in procedure development, but with limited success. Per a DOE order, hourly workers were included on development committees for the lock-out/tag-out procedure and the radiation control procedure. In both cases, hourly workers made suggestions drawing on their field experience to improve the success of procedures once put into place. Their suggestions, however, were ignored. Explaining the order's lack of success on the ground, the worker argues that management basically has a military orientation and that "a private does not talk to a general." The worker argues that this type of DOE order is new at the facility, and it will take time for management to shift from a mode of independent decisionmaking to one that includes the contributions of hourly workers.

In appraising the culture change at the site level, worker contacts identify the new forums for collaboration among labor, contractors, and DOE on issues of the environment, safety, and health as commendable developments. The success of these new structures for incorporating the floor-level perspective suggests the importance of expanded efforts to engage labor in the redirection of the Nuclear Weapons Complex. The tripartite councils, composed of representatives from management, the union, and DOE, have been an effective process for fostering communication among all parties, as well as a mechanism for acting on safety and health suggestions and concerns. Contacts at one facility describe the tripartite council as a format for approaching large-scale problems, rather than minor, day-to-day worker concerns that can be addressed through other avenues. That council is particularly effective because of the high levels of management and DOE staff that participate. The tripartite council also provides an opportunity for labor to move beyond discussions with the contractor and speak directly with DOE in order to address large-scale problems.

The experience of union-appointed health and safety representatives in addressing worker concerns is also suggestive of the benefits of incorporating labor into authority structures. At one facility, for instance, union representatives differ from the professional safety staff in terms of their commitment to investigating worker concerns. These representatives tend to have a better rapport with their coworkers and to be more accessible in the field. Because of their own craft history, union representatives have a personal interest in correcting problems encountered by workers. In addition, workers are more likely to raise concerns with union health and safety representatives than with professional safety staff, because they do not identify with the latter and are aware of both historical and current instances of retribution. This affinity between union-appointed representatives and the work force also exists at another facility.

Through their contributions to standards and procedures development committees, as well as health and safety complaint resolution, workers are beginning to have some authority in how the culture change occurs at the floor level. Lacking in the current environment of change, however, is effective communication between labor and DOE. At one facility, for instance, some improvements in terms of floor-level contributions to the development of procedures are beginning to take place. Nonetheless, labor would like to be included in higher-level decisionmaking processes in order to deal with some of the frustrating aspects of recent changes: continuous oversight by many groups, constant changes in regulations, and rules that make no sense to people on the ground. There are conflicting DOE orders and a variety of new regulations emanating from different oversight groups. Without a uniform way of doing business, explains a labor leader, it is hard to communicate to workers what is expected of them. Workers do not have the opportunity to discuss their experiences with recent DOE orders and to contribute suggestions that can improve the transition. The tripartite councils are a good beginning to partnership between labor and DOE, but the collaboration should not stop there. Labor leaders at all facilities included in this investigation argue for more joint meetings between DOE and worker representatives.

TOO LATE IN COMING FOR SOME

Asking labor leaders to comment on the course of the "culture change" initiated by former Secretary Watkins in 1989 is a request for them to tease out only one aspect of the dramatic changes currently taking place at DOE facilities. Their experience of that culture change is in the context of a very uncertain future for the Nuclear Weapons Complex as a whole. Impending layoffs are likely to affect 20 percent of the work force (4). Dramatic reductions in production requirements are currently taking place, and the landscape of cleanup work is not yet clear. If discussions with labor regarding protections for health and safety shift easily into conversations regarding job security, it is the reflection of a common concern—the exclusion of workers from decisionmaking processes that affect their well-being, both on the shop floor and in general.

To remark on the health and safety aspects of jobs that may disappear is an ill-timed effort, in the eyes of some. In the view of others, however, to understand the contemporary gains or losses in the areas of health and safety, one must recognize what the specter of downsizing can do to a work force. The threat of lost jobs can create an environment antagonistic to a redress of environmental, safety, and health problems. As a worker at one facility explains, during hard economic times such as these when jobs are difficult to find, workers are less likely to raise any complaints about the job. A person fearful of losing his job is not apt to cause trouble—whether in management-labor relations or health and safety issues. This posture is not limited to those with the least power at DOE facilities. A worker at another facility relates that when the plant stopped production in 1985 and there were significant layoffs, everyone—both managers and employees—went into a "survival mode." No one questioned anything.

The engagement of labor in DOE's new culture is particularly important in light of the fear of job loss. Workers must have a sense of investment in the overall changes that are taking place—the necessary reduction of some types of jobs and the overall improvement in the safety and health aspects of those that remain. Labor has already made clear the shared destinies of the workers who held positions under the production-driven regime and those who will hold positions in the new regime. The environment, safety, and health should be a priority within the Weapons Complex, but DOE must also address the adversity that displaced workers will face. Different legislative proposals have been aimed at issues of concern for displaced weapons workers, such as health insurance, medical assistance, and retraining for cleanup work (6).

Empowering the Work Force

Some of the experiences of weapons workers with DOE's new orientation point out lessons for further attempts at creating a truly collaborative culture change. Existing cases of collaboration between management and labor at DOE facilities include the tripartite council, standards and procedures development committees that incorporate administrative and field perspectives, and union-appointed health and safety representatives. These trends should be fostered. The incorporation of labor, however, should not be limited to microcollaboration—a focus simply on the workplace and under the limited care of the contractor. DOE must meet directly, formally, and regularly with workers and their representatives. Furthermore, communication with labor should not be delegated to the lowest levels of authority within the agency.

Other employee empowerment schemes are making their way into various DOE site operations through the efforts of different contractors. One cannot, however, implement an employee involvement program without first conferring with existing worker representatives. The imposition, rather than negotiation, of "employee involvement schemes" will only perpetuate the disempowered position of labor. Management at one facility, for instance, implemented worker involvement schemes to facilitate and address problems associated with mission change. Two of these attempts, "work teams" (performance management teams) and "skills enhancement" (e.g., literacy training), have been unsuccessful, however, in part because management neglected to involve existing worker representatives in the program's development and implementation.

Another facility with a history of good rapport between management and the union has developed a creative solution to the problem of maintenance backlog. Management approached the union and asked if it would be interested in putting together a "SWAT team"—a group of maintenance workers dedicated solely to addressing safety problems normally held up in an overburdened system. The union agreed to the SWAT team, which was, in some respects, a compromise by the union with a long-held tenet of labor—that one does not cross craft lines.

APPENDIX E REFERENCES

1. Advisory Committee on Nuclear Facility Safety to the U.S. Department of Energy, *Final Report on DOE Nuclear Facilities: A Report to the Secretary of Energy* (Washington, DC: U.S. Department of Energy, November 1991), p. 9.
2. Committee to Provide Interim Oversight of the DOE Nuclear Weapons Complex, National Research Council, *The Nuclear Weapons Complex: Management for Health, Safety, and the Environment* (Washington, DC: National Academy Press, 1989), p. 5.
3. Geiger, H. Jack, et al., *Dead Reckoning: A Critical Review of the Department of Energy's Epidemiological Research* (Washington, DC: Physicians for Social Responsibility, 1992), p. 13.
4. Miller, Richard, Atomic Energy Workers Council Meeting, Washington, DC, Mar. 19, 1993.
5. Occupational Safety and Health Administration, *Evaluation of the U.S. Department of Energy's Occupational Safety and Health Program for Its Government-Owned Contractor-Operated Facilities* (Washington, DC: Occupational Safety and Health Administration, December 1990), pp. ii, 47-48.
6. Pollet, Gerald, "Seeking Human Justice," in *Facing Reality: The Future of the U.S. Nuclear Weapons Complex*, Peter Gray (ed.) (Seattle, WA: Nuclear Safety Campaign, May 1992), pp. 31-32.
7. Watkins, James, Secretary, U.S. Department of Energy, "Ten-Point Initiative—Remarks by James D. Watkins, June 27, 1989," reprinted in *Evaluation of the U.S. Department of Energy's Occupational Safety and Health Program for Its Government-Owned Contractor-Operated Facilities* (Washington DC: Occupational Safety and Health Administration, December 1990), pp. B-1 to 13.

Appendix F: Acronyms and Glossary

ABC accelerator-based converter

ACDA Arms Control and Disarmament Agency

AEA Atomic Energy Act

ALARA as low as reasonably achievable

ALMR advanced liquid metal reactor

ATSD (AE) Assistant to the Secretary of Defense for Atomic Energy

CPAF Cost-Plus Award Fee

DNA Defense Nuclear Agency

DNFSB Defense Nuclear Facilities Safety Board

DOD U.S. Department of Defense

DOE U.S. Department of Energy

DP DOE Office of Defense Programs

EA Environmental Assessment

EH DOE Office of Environment, Safety, and Health

EIS Environmental Impact Statement

EM DOE Office of Environmental Restoration and Waste Management

EMTD DOE Executive Management Team for Dismantlement

EPA U.S. Environmental Protection Agency

ERF Energy Research Foundation

FAS Federation of American Scientists

FOIA Freedom of Information Act

fSU former Soviet Union

GESMO Generic Environmental Impact Statement on Mixed Oxide

HE high explosive

HEU highly enriched uranium

HTGR high-temperature gas reactor

IAEA International Atomic Energy Agency

ICBM intercontinental ballistic missile

IFR integral fast reactor

IHE insensitive high explosive

INEL Idaho National Engineering Laboratory

INF Intermediate Range Nuclear Forces Treaty

JAC Joint Advisory Committee on Nuclear Weapons Surety

KUMSC Kirtland Underground Munitions Storage Complex

LANL Los Alamos National Laboratory

LET linear energy transfer

LEU low-enriched uranium

LLNL Lawrence Livermore National Laboratory

LWR light-water reactor

M&H Mason and Hanger-Silas Mason Co., Inc.

M&O management and operations

MHTGR modular HTGR

MINATOM Russian Ministry of Atomic Energy

MIRV multiple independently targetable reentry vehicle

MMES Martin Marietta Energy Systems

MOX mixed oxide

MPN Military Production Network

NE DOE Office of Nuclear Energy

NEPA National Environmental Policy Act

NESSG Nuclear Explosive Safety Study Group

NFS Nuclear Fuel Services

NPT Treaty on the Non-Proliferation of Nuclear Weapons

NRC U.S. Nuclear Regulatory Commission

NRDC	Natural Resources Defense Council
NS	DOE Office of Nuclear Safety
NWSP	Nuclear Weapons Stockpile Plan
OMB	Office of Management and Budget
ORE	Operational Readiness Evaluation
OREPA	Oak Ridge Environmental Peace Alliance
ORR	Operational Readiness Review
OSHA	Occupational Safety and Health Administration
OTA	Office of Technology Assessment
PANAL	Panhandle Area Neighbors and Landowners
PEIS	Programmatic Environmental Impact Statement
PEL	Permissible Exposure Limit
PPE	personal protective equipment
QED	Qualification Evaluation for Dismantlement Release
RADCON	Radiological Control Manual (DOE)
RCRA	Resource Conservation and Recovery Act
RFP	Rocky Flats Plant
SA	DOE Office of Security Affairs
SAC	steel arch construction
SAR	Safety Analysis Report
SLBM	submarine launched ballistic missile
SSD	Safe and Secure Dismantlement Interagency Steering Group
SST	Safe Secure Transport
STAND	Serious Texans Against Nuclear Dumping
STAR	State of Texas Alliance for Resources
START	Strategic Arms Reduction Treaty
STATS	National Academy of Sciences Panel on Separations Technology and Transmutation Systems
SWU	separative work unit
TSA	Technical Safety Appraisal
UCNI	Unclassified Controlled Nuclear Information
USEC	U.S. Enrichment Corporation

Actinides. Radioactive elements with atomic number larger than 88.

Alpha particle. Two neutrons and two protons bound as a single particle emitted from the nucleus of certain radioactive isotopes in the process of decay or disintegration.

Beta particle. A charged particle emitted from the nucleus of certain unstable atomic nuclei (radioactive elements), having the charge and mass of an electron.

Born classified. Term applied to Restricted Data, a category of information deemed "classified from its inception" under the Atomic Energy Act.

Complex 21. Designation for the Nuclear Weapons Complex when current plans for reorganization are realized, expected to occur around the turn of the century.

Criticality. Pertaining to a critical mass (the least amount) of fissionable material that can achieve self-sustaining nuclear chain reactions.

Curie. A unit of radioactivity equal to that emitted by 1 gram of pure radium.

Demilitarization. The process of eliminating or reducing military weapons, materials, other hardware and organizational structures.

Deuterium. An isotope of hydrogen used in the fusion reaction of a nuclear weapon.

Disassembly. The process of taking apart a nuclear warhead and removing the subassemblies, components, and individual parts.

Dismantlement. The process of taking apart a nuclear warhead and removing the subassemplies, components, and individual parts.

Disposition. Determination of the long-term status of materials.

Formerly Restricted Data. Classified information, defined in the Atomic Energy Act, that is shared by DOE and DOD and is related to the military utilization of nuclear weapons or energy. Decisions to declassify such data must be agreed upon by both agencies.

Gamma radiation. Short-wavelength electromagnetic radiation of nuclear origin, similar to. but with higher energy than. x rays.

Gravel Gertie. Term used for explosion-resistant assembly/disassembly bays at the Pantex Plant where nuclear weapons are disassembled.

Half-life. Time required for one-half of the nuclei of a radioactive mass to decay.

High-level waste. Highly radioactive waste material from the reprocessing of spent nuclear fuel (including liquid waste produced directly in reprocessing and any solid waste derived from the liquid) that contains a combination of transuranic waste and fission products in concentrations requiring permanent isolation.

Highly enriched uranium. Uranium enriched in the isotopic content of uranium-235 to greater than 20 percent.

Ionizing radiation. Radiation that separates electrons from an atom or molecule.

Low-enriched uranium. Uranium enriched in the isotopic content of uranium-235, but to less than 20 percent of the total mass.

Low-level waste. Radioactive waste not classified as high-level waste, transuranic waste, spent nuclear fuel, or byproduct material.

National Security Information. Classified information that is not specifically governed by the Atomic Energy Act but by Executive order. The term is used in relation to DOE nuclear defense programs.

Plutonium. Man-made element produced when uranium is irradiated in a reactor. Plutonium-239 is the most suitable isotope for constructing nuclear weapons.

Plutonium pit. The core element of a nuclear weapon's "primary" or fission component. Pits are made of plutonium-239 and surrounded by some type of casing.

Rad. Radiation absorbed dose, a basic unit of absorbed dose of ionizing radiation representing an amount of energy absorbed per unit of absorbing material such as body tissue.

Radionuclide. Certain natural and man-made atomic species with unstable nuclei that can undergo spontaneous breakup or decay and, in the process, emit alpha, beta, or gamma radiation.

Rem (rad equivalent, man). Unit of dose equivalent. The dose equivalent in "rem" is numerically equal to the absorbed dose in "rad" multiplied by necessary modifying factors.

Restricted Data. Classified information defined by the Atomic Energy Act. Restricted Data are born classified, regardless of source.

Secondary. Component of a nuclear weapon that contains elements needed to initiate the fusion reaction in a thermonuclear explosion.

Special nuclear materials. As defined in the Atomic Energy Act, " 'special nuclear materials' means (1) plutonium, uranium enriched in the isotope 233 or in the isotope 235, and any other material which the Commission . . . determines to be special nuclear material, but does not include source material . . .".

Tiger Team. Inspection teams that surveyed DOE Weapons Complex sites pursuant to a June 27, 1989, initiative. Tiger Team inspections gather information for the Secretary of Energy to assess environmental, safety, and health problems at the sites.

Transmutation. A process of converting one element to another by irradiating or bombarding it with radioactive particles.

Transparency. Exchange of information, access to facilities, and cooperative arrangements undertaken to provide ready observation and verification of defense or other activities.

Transuranic. Any element whose atomic number is higher than that of uranium. All transuranic elements are produced artificially and are radioactive.

Tritium. A radioactive gas, an isotope of hydrogen, that serves as a booster for the fusion reaction in the secondary component of a nuclear weapon.

Unclassified Controlled Nuclear Information. Information that is not classified but is judged to be sensitive with respect to DOE defense programs. Its dissemination is therefore controlled and limited. UCNI is a response to a requirement of the Atomic Energy Act.

Vitrification. Process of immobilizing radioactive material by encapsulating it into a glasslike solid.

Warhead. Explosive part of a nuclear weapons system. Warheads consist of nuclear materials, conventional high explosives, and related firing mechanisms.

Weapons retirement. The process by which nuclear weapons are determined to be obsolete or unnecessary for national defense. A retired weapon or weapon system is no longer in an active status or deliverable, but may still be a fully functioning nuclear device.

Appendix G: Reviewers

Steven Aftergood
Federation of American Scientists
Washington, DC

William Bibb
Formerly of the Oak Ridge Operations Office
U.S. Department of Energy
Oak Ridge, TN

Oleg Bukharin
Princeton University
Princeton, NJ

Matthew Bunn
National Academy of Sciences/National Research Council
Washington, DC

Thomas Clements
Greenpeace
Washington, DC

Thomas B. Cochran
Natural Resources Defense Council
Washington, DC

Kevin Cook
Office of the Assistant Secretary of the Army
Washington, DC

Brian Costner
Energy Research Foundation
Columbia, SC

Allen G. Croff
Oak Ridge National Laboratory
Oak Ridge, TN

Paul T. Cunningham
Los Alamos National Laboratory
Los Alamos, NM

Zachary Davis
Congressional Research Service
Washington, DC

Boyd Deaver
Texas Water Commission
Amarillo, TX

Keith H. Dinger
Radiation Health Consultant
Somersworth, NH

Warren H. Donnelly
Congressional Research Service
Washington, DC

James W. Dowie
Office of the Assistant to the Secretary of Defense (Atomic Energy)
Washington, DC

Joel E. Gingold
S.M. Stoller Corp.
Pleasantville, NY

Thomas J. Goodwin
Defense Programs
U.S. Department of Energy
Washington, DC

John M. Googin
Martin Marietta Energy Systems
Oak Ridge, TN

Peter Gray
Institute for Science and International Security
Washington, DC

Charles G. Halsted
U.S. Department of Energy
Germantown, MD

Thomas Johnston
Combustion Engineering and General Atomics
Washington, DC

Steven Krahn
Defense Nuclear Facilities Safety Board
Washington, DC

Steve Lipmann
Environmental Research and Analysis
Irvington, NY

Gregory Lyle
Defense Nuclear Agency
Washington, DC

Arjun Makhijani
Institute for Energy and Environmental Research
Takoma Park, MD

Joseph Martillotti
Texas Department of Health
Bureau of Radiation Control
Austin, TX

Jon Medalia
Congressional Research Service
Washington, DC

Marvin Miller
Massachusetts Institute of Technology
Cambridge, MA

Loring Mills
Formerly of Edison Electric Institute
Washington, DC

Bill Mitchell
Nuclear Safety Campaign
Seattle, WA

Robert Morgan
Washington University
St. Louis, MO

Joseph Pappe
U.S. Air Force
Albuquerque, NM

Arvin S. Quist
Oak Ridge National Laboratory
Oak Ridge, TN

Paul Shutt
Nuclear Fuel Services
Erwin, TN

Lynn Stembridge
Hanford Education Action League
Spokane, WA

William G. Sutcliffe
Lawrence Livermore National Laboratory
Livermore, CA

John J. Taylor
Electric Power Research Institute
Palo Alto, CA

Pierre Toureille
General Accounting Office
Washington, DC

Troy Wade II
Wade Associates (formerly the Assistant Secretary of Energy, Defense Programs)
Washington, DC

☆ U.S. GOVERNMENT PRINTING OFFICE: 1993 - 343-283 - 814/96955

Superintendent of Documents **Publications** Order Form

Order Processing Code:
*7170

P3

Telephone orders (202) 783-3238
To fax your orders (202) 512-2250
Charge your order. It's Easy!

☐ **YES**, please send me the following:

_______ copies of ***Dismantling the Bomb and Managing the Nuclear Materials (212 pages),*** S/N 052-003-01343-6 at $12.00 each.

The total cost of my order is $_________. International customers please add 25%. Prices include regular domestic postage and handling and are subject to change.

(Company or Personal Name) (Please type or print)

(Additional address/attention line)

(Street address)

(City, State, ZIP Code)

(Daytime phone including area code)

(Purchase Order No.)

Please Choose Method of Payment:

☐ Check Payable to the Superintendent of Documents

☐ GPO Deposit Account ☐☐☐☐☐☐☐ — ☐

☐ VISA or MasterCard Account

☐☐☐☐☐☐☐☐☐☐☐☐☐☐☐☐☐☐☐☐

☐☐☐☐ (Credit card expiration date)

Thank you for your order!

(Authorizing Signature) (9/93)

May we make your name/address available to other mailers? **YES** ☐ **NO** ☐

Mail To: New Orders, Superintendent of Documents, P.O. Box 371954, Pittsburgh, PA 15250-7954

THIS FORM MAY BE PHOTOCOPIED

Superintendent of Documents **Publications** Order Form

Order Processing Code:
*7170

P3

Telephone orders (202) 783-3238
To fax your orders (202) 512-2250
Charge your order. It's Easy!

☐ **YES**, please send me the following:

_______ copies of ***Dismantling the Bomb and Managing the Nuclear Materials (212 pages),*** S/N 052-003-01343-6 at $12.00 each.

The total cost of my order is $_________. International customers please add 25%. Prices include regular domestic postage and handling and are subject to change.

(Company or Personal Name) (Please type or print)

(Additional address/attention line)

(Street address)

(City, State, ZIP Code)

(Daytime phone including area code)

(Purchase Order No.)

Please Choose Method of Payment:

☐ Check Payable to the Superintendent of Documents

☐ GPO Deposit Account ☐☐☐☐☐☐☐ — ☐

☐ VISA or MasterCard Account

☐☐☐☐☐☐☐☐☐☐☐☐☐☐☐☐☐☐☐☐

☐☐☐☐ (Credit card expiration date)

Thank you for your order!

(Authorizing Signature) (9/93)

May we make your name/address available to other mailers? **YES** ☐ **NO** ☐

Mail To: New Orders, Superintendent of Documents, P.O. Box 371954, Pittsburgh, PA 15250-7954

THIS FORM MAY BE PHOTOCOPIED